Math Mammoth
Addition 1

By Maria Miller

Contents

Introduction 5

Games and Activities 6

Two Groups and a Total 9

Learn the Symbols + and = ... 12

Addition Practice 1 15

Which is More? 17

Missing Items 19

Sums with 5 24

Sums with 6 26

Adding on a Number Line 28

Sums with 7 32

Sums with 8 35

Adding Many Numbers 38

Addition Practice 2 41

Sums with 9 43

Sums with 10 47

Comparisons 51

Review of Addition Facts 54

Answers 59

More from Math Mammoth 69

Introduction

Math Mammoth Addition 1 is a self-explanatory worktext, dealing with the concept of addition and addition facts within 0-10 (in few occasions numbers between 10 and 20 are used). It is most suitable for kindergarten and first grade.

The book starts out with very easy and visual addition problems within 0-5, where children can simply count the objects to add. You can easily adapt these early lessons to be done with concrete objects or manipulatives.

If the child does not know the symbols " + " and " = " yet, you can introduce them *orally* at first. Use blocks or other objects to illustrate, and say: "Three blocks and four blocks makes seven blocks. Three blocks *plus* four blocks *equals* seven blocks." Then ask the child to make an addition problem with the objects, using those words. Play like that until the child can use the words "plus" and "equals" in their own speech. This will make it easier to learn to use the written symbols.

In the lesson *Which Is More?*, the symbols " < " and " > " are introduced as being like a "hungry alligator's mouth." In this lesson, children only compare numbers, such as $5 < 7$. In later lessons, children will also learn to compare expressions, such as $2 + 3 < 4 + 4$.

Next, we introduce "missing addend" problems, or problems such as $1 + \_\_\_ = 5$. First, the lesson uses pictures, and then gradually only symbols. These problems are very important, as they lead the child to learn the connection between addition and subtraction.

A child might confuse the missing number problem $1 + \_\_\_ = 5$ with $1 + 5 = \_\_\_$. To help the child see the difference, word these problems like this: "One and *how many more* make five?"

You can also model missing addend problems by drawing. In our example of $1 + \_\_\_ = 5$, the teacher would first draw one stick, and then tell the student, "We need a total of five sticks. Draw more until there are five of them." The number of sticks that the child needs to draw in order to make five is the number that goes on the empty line. So, you can say, "First there was one stick, then you needed to add (draw) some more to make 5. How many more did you draw?"

Then we come to the lesson *Sums with 5*. It practices the number bonds (number combinations) that add up to 5, which are 0 and 5, 1 and 4, and 2 and 3. After that, we study sums with 6, sums with 7, and so on. The goal of these lessons is to help the child to memorize addition facts within 10. However, the child does not need to fully memorize them yet. All of these lessons are building toward that goal, but the final mastery of addition facts does not have to happen this early in first grade.

My approach to memorizing the basic addition facts within 0-10 is many-fold:

1. Structured drills, such as used in the lessons *Sums with 5*, *Sums with 6*, and so on, are not random drills, because they use the pattern or the structure in the facts. This will connect the facts to a context, and help the child to better understand the facts on a conceptual level, instead of merely memorizing them at random. In each of these lessons, the child learns the number combinations that add up to the specific number (aka number bonds). This understanding is the basis for the drills.

2. Using addition facts in games and in everyday life is very helpful — and especially in games, because most children like to play games.

3. Random drilling may also be used, sparingly, as one tool among others.

4. Memory helpers can be silly mnemonics or writing math facts on a poster and hanging it on the wall. Not all children need these, but feel free to use them if you like.

Another important thread running through the book is to develop a child's understanding of the symbols $+$, $<$, and $>$. The lessons in the book help children get used to equations such as $9 = 5 + 4$ and inequalities such as $2 < 5 + 4$, to build the correct understanding of these symbols.

Many children develop a misconception of the equals sign being an "operator," as if it means that you need to add/subtract/multiply/divide, or "operate" on the numbers in the equation. A child with this misconception will treat the equation 9 = __ + 4 as an addition problem 9 + 4. Instead, the equals sign is just that, signifying that what is on the right and left side of the sign are equal in value.

A number line is an important way to model addition, as it helps to build number sense and ties in with measurement. Children also encounter addition tables, number patterns, word problems, and get used to a symbol for the unknown number (such as in ☐ + 5 = 10). So, while it may look on the surface that all we do is add small numbers, actually, a lot happens in this book!

Please also see the following pages for games that I recommend while studying this book. Games are important at this level, as they help children to practice the addition facts and also make math fun.

Lastly, don't forget to check out the free videos matched to the curriculum at https://www.mathmammoth.com/videos/.

Games and Activities

Some Went Hiding

You need: The same number of small objects as the sum you are studying. For example, to study the sums with 5, you need 5 objects (marbles, blocks, etc).

Game play: The first player shows the objects but quickly hides some of them behind their back without showing how many. Then they show the remaining objects to the next player, who has to say how many "went hiding." If the player gives the right answer, it is then their turn to hide some and ask the next player to answer. If they give a wrong answer, they forfeit their turn. This game appeals best to young children.

Variation: Instead of getting a turn to hide objects, the player who answers correctly may gain points or other rewards for the right answer.

10 Out (or *6 Out, 7 Out, 8 Out, etc.*)

You need: A deck of number cards with numbers 1-10, or regular playing cards without the face cards.

Preparation: Choose a target sum, such as 10. Deal seven cards to each player. Place the rest face down in a pile in the middle of the table.

Game play: At your turn, first take one card from the pile. Then try to find pairs of cards in your hand that add up to 10, and discard any such pairs. Discard the card 10 also if you have it. If you cannot find any such pairs, ask for any one card you want (such as 6) from the player to your right (as in "Go Fish"). That player, if they have it, must give it, and you will then discard the pair that makes 10. Then it is the next player's turn. The player who first discards all the cards from their hand is the winner.

Variations:
* Deal more than seven cards.
* Deal fewer cards if there are a lot of players or the players are very young.
* Allow players to discard *three* cards that add up to 10.
* Instead of ten, players discard cards that add up to 5, 6, 7, 8, or 9.

Number Bonds in the Pond

You need: A standard deck (or several) of playing cards or number cards

Preparation: Choose a target sum for the game. If the target sum is 5, make a deck of cards consisting of numbers 1 through 4. If the target sum is 6, make a deck of numbers 1-5. And so on. (The deck always consists of numbers that are from 1 through X − 1 where X is the target sum.) Place a target number card face up between the players, and spread out the rest of the cards face down, like a pond, between the players.

Game play: At your turn, if you don't have any cards in your hand, take <u>two</u> cards from the pond. If you do, take <u>one</u> card from the pond. Now check if any two cards in your hand add up to the target number. If so, put those cards away to your personal pile. If not, it is the next player's turn. The game ends when there are no more cards in the pond. The winner is the person with most cards in their personal pile.

Variation: Allow three cards/numbers to be added to reach the target number.

Notes: Depending on the number of players, you may need several decks of cards to make the pond. When first playing this game with your child/students, start with 5 as the target number, and advance to target numbers of 6, 7, and onward. Playing this game several times will help the child to memorize the number bonds (the sums) associated with a particular target number.

Addition Challenge

You need: A standard deck of playing cards from which you remove the face cards and perhaps also some of the other higher-numbered cards, such as tens, nines, and eights. Alternatively, a set of dominoes works well for children who do not yet know the sums beyond 12.

Game Play: In each round, each player is dealt two cards face up, and calculates the sum. The player with the highest sum gets all the cards from the other players. After enough rounds have been played to use all of the cards, the player with the most cards wins. If two or more players have the same sum, those players get an additional two cards and use those to resolve the tie.

Any **board game** where you move the piece by rolling two dice also works well to practice addition.

Games and Activities at Math Mammoth Practice Zone

Single-Digit Addition Practice
https://www.mathmammoth.com/practice/addition-single-digit

Addition Hidden Picture Game
https://www.mathmammoth.com/practice/mystery-picture

Number Bonds
https://www.mathmammoth.com/practice/number-bonds

"7 Up" Addition Facts
https://www.mathmammoth.com/practice/seven-up

Fruity Math
https://www.mathmammoth.com/practice/fruity-math

Further Resources on the Internet

We have compiled a list of external Internet resources that match the topics in this book. This list of links includes web pages that offer:

- **online practice** for concepts;

- online **games**, or occasionally, printable games;

- **animations** and interactive **illustrations** of math concepts;

- **articles** that teach a math concept.

We heartily recommend you take a look at the list. Many of our customers love using these resources to supplement the bookwork. You can use the resources as you see fit for extra practice, to illustrate a concept better, and even just for some fun. Enjoy!

https://l.mathmammoth.com/blue/addition1

Scan me

Two Groups and a Total

1. Make two groups.

a. 4 1 and 3	**b.** 4 2 and 2	**c.** 4 3 and 1
d. 5 3 and 2	**e.** 5 2 and 3	**f.** 5 1 and 4

2. Make two groups. Write how many are in the second group.

a. 4 1 and _____	**b.** 4 2 and _____	**c.** 4 3 and _____
d. 5 4 and _____	**e.** 5 3 and _____	**f.** 5 2 and _____
g. 5 1 and _____	**h.** 5 5 and _____	**i.** 5 0 and _____

3. Draw as many dots as the number shows. Then divide them into two groups.
   (There are many ways to do this.) Write how many are in each group.

a. 3	b. 5	c. 4
_____ and _____	_____ and _____	_____ and _____
d. 2	e. 6	f. 8
_____ and _____	_____ and _____	_____ and _____

4. The number at the top is the total. Draw the missing dots on the face of the blank dice.
   Write on the lines how many dots are on the face of each dice.

a. 3	b. 6	c. 5
_____ and _____	_____ and _____	_____ and _____
d. 4	e. 6	f. 5
_____ and _____	_____ and _____	_____ and _____

2 and 2   4

"Two and two makes four."

1 and 4   5

"One and four makes five."

5. Write how many are in each group. Write the total in the box.

a.

_____ and _____

b.

_____ and _____

c.

_____ and _____

d.

_____ and _____

e.

_____ and _____

f.

_____ and _____

g.

_____ and _____

h.

_____ and _____

i.

_____ and _____

6. Draw circles for each number. Write the total in the box.

a. 2 and 2

b. 3 and 1

c. 3 and 3

d. 1 and 4

11

Learn the Symbols + and =

$$3 + 2 = 5$$

THREE plus TWO equals FIVE

$$1 + 3 = 4$$

ONE plus THREE equals FOUR

1. Fill in the numbers. Add. Read the additions aloud using "plus" and "equals".

a. $\underline{1} + \underline{3} = $	b. $\underline{} + \underline{} = $
c. $\underline{} + \underline{} = $	d. $\underline{} + \underline{} = $
e. $\underline{} + \underline{} = $	f. $\underline{} + \underline{} = $
g. $\underline{} + \underline{} = $	h. $\underline{} + \underline{} = $
i. $\underline{} + \underline{} = $	j. $\underline{} + \underline{} = $

2. Write the numbers. Add. Read the additions aloud using "plus" and "equals".

a.

_____ + _____ =

b.

_____ + _____ =

c.

_____ + _____ =

d.

_____ + _____ =

3. Add with zero.

a.

$\underline{2} + \underline{0} = $ _____

b.

_____ + _____ = _____

c.

_____ + _____ = _____

d.

_____ + _____ = _____

e.

_____ + _____ = _____

f.

_____ + _____ = _____

g.

_____ + _____ = _____

h.

_____ + _____ = _____

4. Write how many dots. Then add.

a.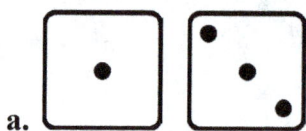

_____ + _____ = _____

b.

_____ + _____ = _____

c.

_____ + _____ = _____

d.

_____ + _____ = _____

e.

_____ + _____ = _____

f.

_____ + _____ = _____

g.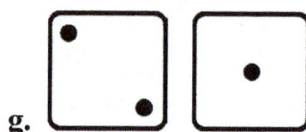

_____ + _____ = _____

h.

_____ + _____ = _____

i.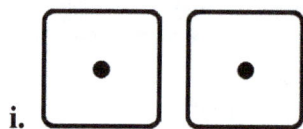

_____ + _____ = _____

j.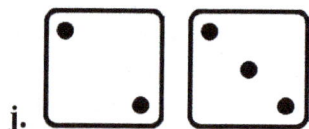

_____ + _____ = _____

Addition Practice 1

1. In the second box, draw enough things to show the second number. Then add.

a. $2 + 1 =$ _____

b. $3 + 2 =$ _____

c. $1 + 2 =$ _____

d. $4 + 1 =$ _____

e. $2 + 3 =$ _____

f. $0 + 4 =$ _____

g. $2 + 2 =$ _____

h. $1 + 0 =$ _____

i. $3 + 1 =$ _____

2. Draw dots in each box for the numbers. Then add.

a. $2 + 2 =$ _____

b. $1 + 3 =$ _____

c. $0 + 5 =$ _____

d. $4 + 1 =$ _____

e. $2 + 3 =$ _____

f. $1 + 3 =$ _____

3. Add. If you want to, you can draw balls or sticks to help you.

a. $1 + 2 = $ _____	b. $3 + 0 = $ _____	c. $2 + 2 = $ _____
d. $2 + 3 = $ _____	e. $1 + 4 = $ _____	f. $0 + 5 = $ _____
g. $3 + 2 = $ _____	h. $2 + 1 = $ _____	i. $4 + 1 = $ _____

4. Add in both orders! Notice: the answer is the same. You can draw marbles to help.

a. $2 + 3 = 5$ $3 + 2 = 5$	b. $1 + 2 = $ _____ $2 + 1 = $ _____	c. $3 + 1 = $ _____ $1 + 3 = $ _____
d. $1 + 4 = $ _____ $4 + 1 = $ _____	e. $0 + 2 = $ _____ $2 + 0 = $ _____	f. $5 + 0 = $ _____ $0 + 5 = $ _____

16

Which is More?

The symbols < and > are like a "hungry alligator's mouth."
The mouth always opens towards the **bigger** number.

1 < 4

One <u>is less than</u> four.

< means "IS LESS THAN".

5 > 3

Five <u>is greater than</u> three.

> means "IS GREATER THAN".

1. Practice writing < and >.

2. Circle the bigger number. Read using "less than" and "greater than".

a. 1 < 3

"1 is less than 3".

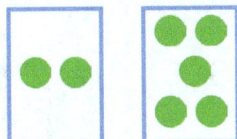

b. 2 < 5

"2 is less than 5".

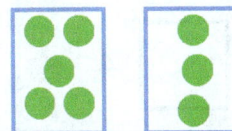

c. 5 > 3

"5 is greater than 3".

d. 6 > 2

e. 4 > 1

f. 2 < 4

3. Circle the bigger number. Read using "less than" and "greater than".

a. 6 > 0	**b.** 3 < 4	**c.** 4 < 5	**d.** 4 > 3
e. 1 < 2	**f.** 2 > 1	**g.** 3 < 5	**h.** 0 < 4

4. Write < or > in the box.

a. 1 ☐ 4

b. 2 ☐ 5

c. 6 ☐ 3

d. 3 ☐ 4

e. 5 ☐ 1

f. 2 ☐ 3

5. Write < or > between the numbers. You can draw circles to help you.

a. 1 ☐ 4

b. 4 ☐ 3

c. 2 ☐ 5

d. 0 ☐ 4

6. Write < or > between the two numbers.

a. 1 ☐ 4

b. 4 ☐ 5

c. 2 ☐ 4

d. 5 ☐ 3

e. 1 ☐ 2

f. 3 ☐ 1

g. 5 ☐ 4

h. 4 ☐ 6

i. 3 ☐ 5

j. 1 ☐ 0

k. 2 ☐ 5

l. 0 ☐ 2

Missing Items

Something is missing from the addition.
The TOTAL is not missing. The total is 5.

How many are in the second group? That is what is missing!

There should be a total of 5 dots. Draw 4 in the face of the second dice.

5

$\boxed{\,\cdot\,}$ + $\boxed{\ }$

1 + _____

There should be a total of 4 dots. The face of the second dice has two. There are none on the face of the first dice, so you need to draw them.

Read: "2 plus what number makes 4?"
  or, "2 and how many more makes 4?"
  or, "What number and 2 makes 4?"

4

$\boxed{\ }$ + $\boxed{\,\cdot\ \cdot\,}$

_____ + 2

1. Complete the addition. Draw the missing dots. The total is on top.

3	3	5
$\boxed{\,\cdot\,}$ + $\boxed{\ }$	$\boxed{\,\cdot\ \cdot\,}$ + $\boxed{\ }$	$\boxed{\ }$ + $\boxed{::}$
a. 1 + _____	b. 2 + _____	c. _____ + 4
5	5	4
$\boxed{\therefore}$ + $\boxed{\ }$	$\boxed{\ }$ + $\boxed{\,\cdot\ \cdot\,}$	$\boxed{\ }$ + $\boxed{\therefore}$
d. 3 + _____	e. _____ + 2	f. _____ + 3
5	4	4
$\boxed{\vdots\,\cdot}$ + $\boxed{\ }$	$\boxed{\ }$ + $\boxed{\,\cdot\,}$	$\boxed{\ }$ + $\boxed{\,\cdot\ \cdot\,}$
g. 5 + _____	h. _____ + 1	i. _____ + 2

$3 + \underline{} = \boxed{5}$

$\underline{} + 3 = \boxed{4}$

The **TOTAL** is now written after the equal sign " = ".
The answer is $3 + \underline{2} = \boxed{5}$

See the **TOTAL** written after the equal sign " = ".
The answer is $\underline{1} + 3 = \boxed{4}$

2. Draw more dots to show the missing number. Write the missing number.

a. $2 + \underline{} = 4$

b. $1 + \underline{} = 1$

c. $\underline{} + 1 = 5$

d. $3 + \underline{} = 5$

e. $\underline{} + 1 = 4$

f. $2 + \underline{} = 3$

g. $5 + \underline{} = 5$

h. $\underline{} + 1 = 3$

i. $2 + \underline{} = 5$

j. $1 + \underline{} = 5$

k. $\underline{} + 2 = 2$

l. $3 + \underline{} = 4$

3. Draw dots in the empty box for the missing number. Read the problems aloud:
   "2 plus how many makes 4?"

a. $2 + \underline{\quad} = 4$	b. $4 + \underline{\quad} = 4$	c. $\underline{\quad} + 3 = 4$
d. $2 + \underline{\quad} = 5$	e. $\underline{\quad} + 1 = 3$	f. $\underline{\quad} + 4 = 5$

There are no dots on the face of either dice.

The face of the first dice is missing its dots. The face of the second is *supposed* to have none, since there is a zero below it.

Draw 4 dots on the face of the first dice, because $\underline{4} + 0 = 4$.

$\underline{\quad} + 0 = $ 4

4. Draw dots in the boxes for the missing numbers. Notice that some boxes are supposed to have zero dots.

a. $0 + \underline{\quad} = 4$	b. $2 + \underline{\quad} = 2$	c. $\underline{\quad} + 0 = 3$
d. $3 + \underline{\quad} = 3$	e. $\underline{\quad} + 0 = 2$	f. $0 + \underline{\quad} = 1$

21

5. Draw dots to illustrate each addition problem. Find what number is missing.

a. 4 + _____ = 5

b. 1 + _____ = 2

c. _____ + 3 = 5

d. 3 + _____ = 4

e. 2 + _____ = 3

f. _____ + 2 = 4

g. 1 + _____ = 5

h. _____ + 1 = 4

i. 3 + _____ = 3

6. Solve. Now, the missing number goes inside the shape. You can draw dots to help you. Remember, the number after the " = " sign is the total.

a. 2 + [ ] = 5

b. [ ] + 2 = 4

c. [ ] + 1 = 3

d. [ ] + 3 = 3

e. 3 + [ ] = 5

f. 0 + [ ] = 2

g. 3 + [ ] = 4

h. [ ] + 2 = 4

i. [ ] + 1 = 5

7. Practice "normal" addition.

a. $1 + 1 =$ _____

$2 + 1 =$ _____

b. $4 + 0 =$ _____

$3 + 1 =$ _____

c. $1 + 4 =$ _____

$2 + 2 =$ _____

d. $2 + 3 =$ _____

$1 + 4 =$ _____

e. $0 + 5 =$ _____

$1 + 2 =$ _____

f. $3 + 2 =$ _____

$4 + 1 =$ _____

8. Find the missing number. The marbles illustrate the total. Notice the patterns!

a.

$0 +$ _____ $= 3$

$1 +$ _____ $= 3$

$2 +$ _____ $= 3$

$3 +$ _____ $= 3$

b.

$0 +$ _____ $= 4$

$1 +$ _____ $= 4$

$2 +$ _____ $= 4$

$3 +$ _____ $= 4$

$4 +$ _____ $= 4$

c.

$0 +$ _____ $= 5$

$1 +$ _____ $= 5$

$2 +$ _____ $= 5$

$3 +$ _____ $= 5$

$4 +$ _____ $= 5$

$5 +$ _____ $= 5$

Sums with 5

1. Here are some different ways to group five elephants into two groups. The "|" symbol separates the two groups. Write the addition sentences.

_____ + _____ = _____   _____ + _____ = _____

_____ + _____ = _____   _____ + _____ = _____

_____ + _____ = _____   _____ + _____ = _____

2. Add.

a. $4 + 1 =$ _____

$2 + 2 =$ _____

$3 + 2 =$ _____

$1 + 2 =$ _____

b. $2 + 3 =$ _____

$1 + 3 =$ _____

$1 + 4 =$ _____

$2 + 1 =$ _____

c. $1 + 1 =$ _____

$0 + 5 =$ _____

$1 + 4 =$ _____

$3 + 2 =$ _____

3. Play "5 Out" *and/or* "Some Went Hiding" with 5 objects (see the introduction).

4. **Drill.** Don't write the answers in the boxes, but just solve them in your head.

$1 + \boxed{} = 5$ $\qquad$ $4 + \boxed{} = 5$ $\qquad$ $\boxed{} + 2 = 5$ $\qquad$ $\boxed{} + 3 = 5$

$2 + \boxed{} = 5$ $\qquad$ $3 + \boxed{} = 5$ $\qquad$ $\boxed{} + 0 = 5$ $\qquad$ $\boxed{} + 1 = 5$

$0 + \boxed{} = 5$ $\qquad$ $5 + \boxed{} = 5$ $\qquad$ $\boxed{} + 4 = 5$ $\qquad$ $\boxed{} + 5 = 5$

5. Add. Compare the problems in each group. You can draw more shapes to help you with the additions.

a. $4 + 0 =$ _____

$4 + 1 =$ _____

$4 + 2 =$ _____

b. $6 + 0 =$ _____

$6 + 1 =$ _____

$6 + 2 =$ _____

c. $3 + 0 =$ _____

$3 + 1 =$ _____

$3 + 2 =$ _____

d. $7 + 0 =$ _____

$7 + 1 =$ _____

$7 + 2 =$ _____

e. $5 + 0 =$ _____

$5 + 1 =$ _____

$5 + 2 =$ _____

f. $8 + 0 =$ _____

$8 + 1 =$ _____

$8 + 2 =$ _____

6. Draw more things to illustrate the missing number. Complete the addition sentence.

a. __2__ + _____ = 5

b. _____ + _____ = 3

c. _____ + _____ = 4

d. _____ + _____ = 5

e. _____ + _____ = 6

f. _____ + _____ = 6

25

Sums with 6

1. Here are some different ways to group six hippos into two groups.
   Write the addition sentences.

_____ + _____ = _____   _____ + _____ = _____

_____ + _____ = _____   _____ + _____ = _____

_____ + _____ = _____   _____ + _____ = _____

_____ + _____ = _____

2. Play "6 Out" *and/or* "Some Went Hiding" with 6 objects (see the introduction).

3. **Drill.** Don't write the answers but just solve them in your head.

$1 + \boxed{} = 6$ $4 + \boxed{} = 6$ $\boxed{} + 2 = 6$ $\boxed{} + 3 = 6$

$2 + \boxed{} = 6$ $3 + \boxed{} = 6$ $\boxed{} + 0 = 6$ $\boxed{} + 1 = 6$

$6 + \boxed{} = 6$ $5 + \boxed{} = 6$ $\boxed{} + 4 = 6$ $\boxed{} + 5 = 6$

4. Add the numbers and write the total on the line.

a. $1 + 5 = $ _____	b. $2 + 3 = $ _____	c. $4 + 2 = $ _____

5. Draw more little boxes to illustrate the missing number.

a. $2 + \boxed{} = 6$

b. $2 + \boxed{} = 5$

c. $4 + \boxed{} = 6$

d. $3 + \boxed{} = 6$

e. $1 + \boxed{} = 6$

f. $5 + \boxed{} = 6$

g. $1 + \boxed{} = 5$

h. $0 + \boxed{} = 6$

i. $3 + \boxed{} = 5$

6. Jack and Jill share 5 cucumbers and 6 lemons in different ways. Find how many Jill gets. You can cover the cucumbers or lemons with your hand to help.

a. 5

Jack gets:	Left for Jill:
2	
1	
5	
3	
0	
4	

b. 6

Jack gets:	Left for Jill:
1	
4	
5	
0	
2	
3	

7. Add.

$2 + 3 = \underline{}$

$4 + 1 = \underline{}$

$3 + 3 = \underline{}$

$4 + 2 = \underline{}$

$1 + 3 = \underline{}$

$1 + 5 = \underline{}$

$2 + 2 = \underline{}$

$2 + 4 = \underline{}$

Adding on a Number Line

$$5 + 3 = 8$$

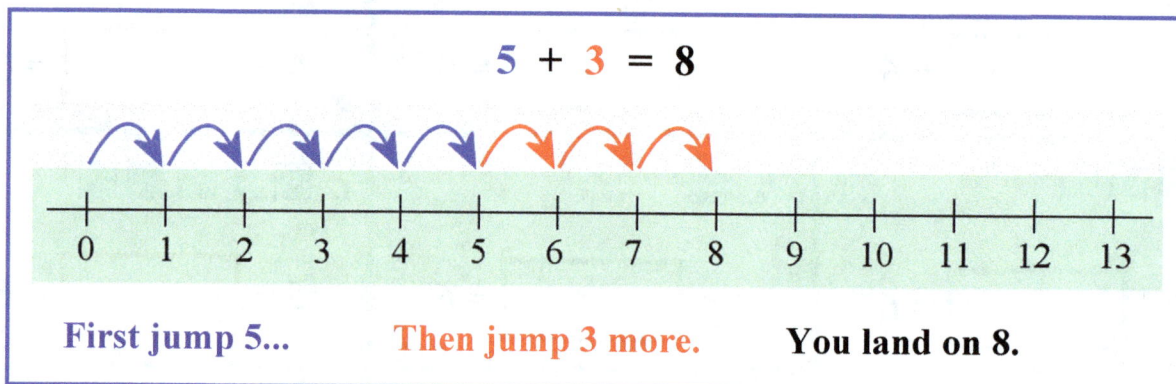

First jump 5...   **Then jump 3 more.**   **You land on 8.**

1. Draw the jumps to illustrate the addition and find the answer.
   You can use a different color for each number when you draw the jumps.

a. $5 + 2 = $ _____

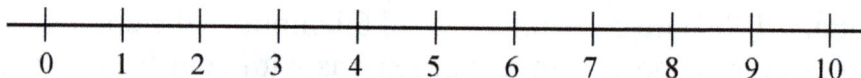

b. $4 + 1 = $ _____

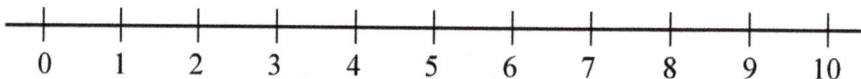

c. $6 + 3 = $ _____

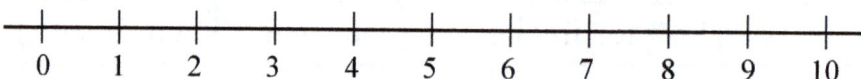

d. $9 + 1 = $ _____

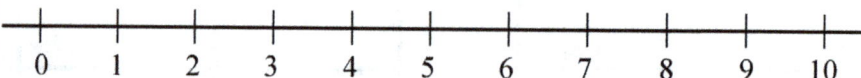

e. $7 + 3 = $ _____

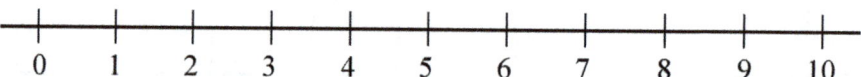

f. $4 + 3 = $ _____

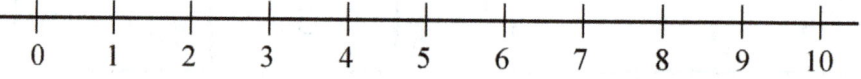

28

OR.... $5 + 3 = 8$

| 0 | 1 | 2 | 3 | 4 | 5 | 6 | 7 | 8 | 9 | 10 | 11 | 12 | 13 |

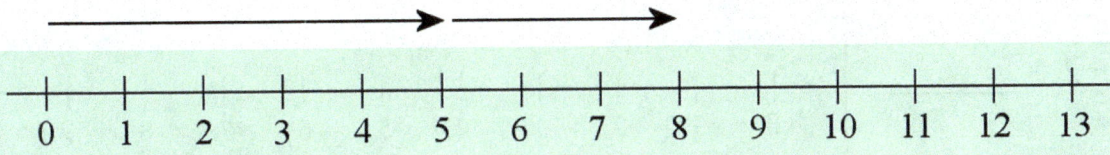

First draw an arrow that is 5 long. Then draw an arrow that is 3 long.

2. Write the addition sentence.

a.

| 0 | 1 | 2 | 3 | 4 | 5 | 6 | 7 | 8 | 9 | 10 |

_____ + _____ = _____

b.

| 0 | 1 | 2 | 3 | 4 | 5 | 6 | 7 | 8 | 9 | 10 |

_____ + _____ = _____

c.

| 0 | 1 | 2 | 3 | 4 | 5 | 6 | 7 | 8 | 9 | 10 |

_____ + _____ = _____

d.

| 0 | 1 | 2 | 3 | 4 | 5 | 6 | 7 | 8 | 9 | 10 |

_____ + _____ = _____

e.

| 0 | 1 | 2 | 3 | 4 | 5 | 6 | 7 | 8 | 9 | 10 |

_____ + _____ = _____

f.

| 0 | 1 | 2 | 3 | 4 | 5 | 6 | 7 | 8 | 9 | 10 |

_____ + _____ = _____

g.

| 0 | 1 | 2 | 3 | 4 | 5 | 6 | 7 | 8 | 9 | 10 |

_____ + _____ = _____

3. Draw arrows (or jumps) to show the addition.

a. $7 + 2 =$ _____

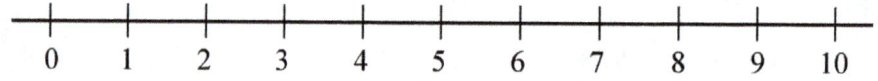

b. $2 + 3 =$ _____

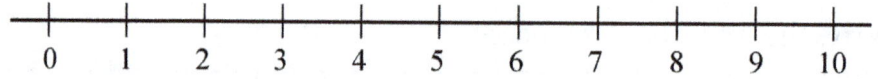

c. $5 + 4 =$ _____

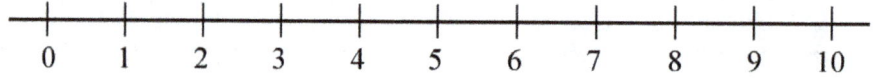

d. $7 + 1 =$ _____

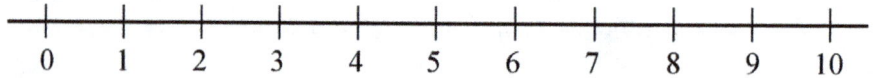

e. $6 + 4 =$ _____

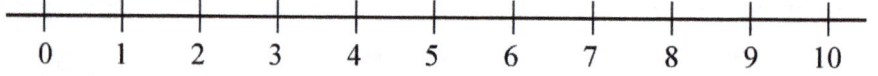

f. $4 + 2 =$ _____

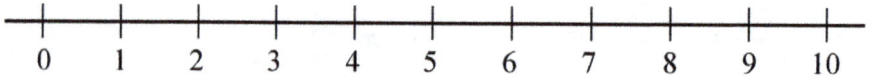

g. $9 + 1 =$ _____

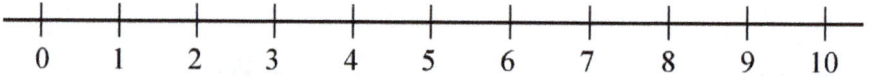

h. $1 + 4 =$ _____

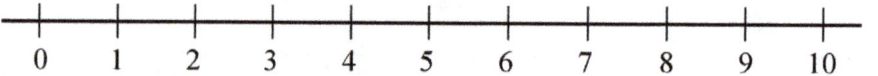

4. Write the addition sentence for each picture.
   If the child is not familiar with numbers greater than 10, you can skip these.

a. _____ + _____ = _____

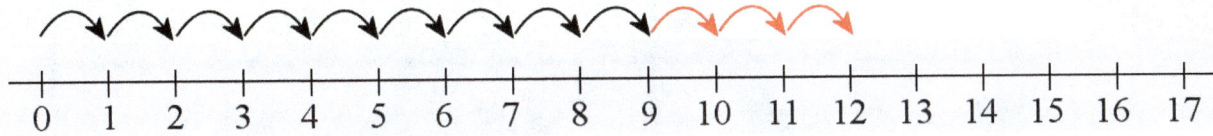

b. _____ + _____ = _____

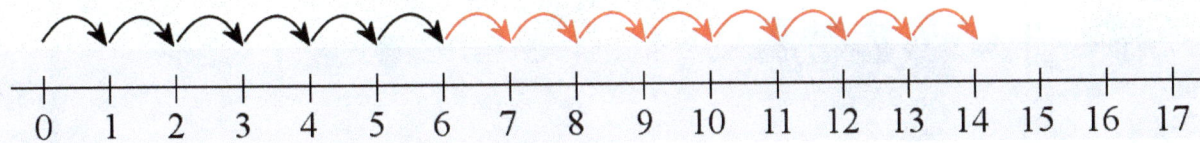

c. _____ + _____ = _____

5. Add "1", add "2" to the number. Use the number line to help.

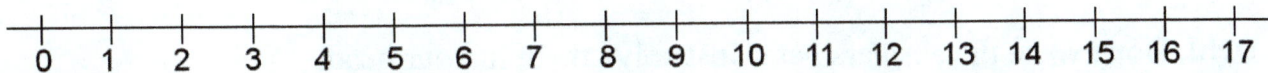

a.	b.	c.	d.
$7 + 1 =$ _____	$5 + 1 =$ _____	$6 + 1 =$ _____	$8 + 1 =$ _____
$7 + 2 =$ _____	$5 + 2 =$ _____	$6 + 2 =$ _____	$8 + 2 =$ _____

e.	f.	g.	h.
$10 + 1 =$ _____	$12 + 1 =$ _____	$13 + 1 =$ _____	$11 + 1 =$ _____
$10 + 2 =$ _____	$12 + 2 =$ _____	$13 + 2 =$ _____	$11 + 2 =$ _____

Sums with 7

1. Here are some different ways to group seven marbles into two groups. Write the addition sentences.

_____ + _____ = _____   _____ + _____ = _____

_____ + _____ = _____   _____ + _____ = _____

_____ + _____ = _____   _____ + _____ = _____

_____ + _____ = _____   _____ + _____ = _____

2. **Drill.** Don't write the answers here. Just solve them in your head.

$5 + \square = 7$ $2 + \square = 7$ $6 + \square = 7$ $\square + 3 = 7$ $\square + 7 = 7$

$3 + \square = 7$ $1 + \square = 7$ $0 + \square = 7$ $\square + 2 = 7$ $\square + 1 = 7$

$7 + \square = 7$ $4 + \square = 7$ $4 + \square = 7$ $\square + 6 = 7$ $\square + 5 = 7$

3. Add.

a.	b.	c.	d.
$3 + 3 =$ _____	$5 + 2 =$ _____	$6 + 1 =$ _____	$2 + 5 =$ _____
$3 + 4 =$ _____	$4 + 2 =$ _____	$4 + 3 =$ _____	$4 + 2 =$ _____

4. Play "7 Out" *and/or* "Some Went Hiding" with 7 objects (see the introduction).

5. Fill in the missing numbers. You may draw dots to help. Notice the patterns!

a.	b.	c.
$0 + \underline{\hspace{1.5cm}} = 7$	$0 + \underline{\hspace{1.5cm}} = 6$	$0 + \underline{\hspace{1.5cm}} = 5$
$1 + \underline{\hspace{1.5cm}} = 7$	$1 + \underline{\hspace{1.5cm}} = 6$	$1 + \underline{\hspace{1.5cm}} = 5$
$2 + \underline{\hspace{1.5cm}} = 7$	$2 + \underline{\hspace{1.5cm}} = 6$	$2 + \underline{\hspace{1.5cm}} = 5$
$3 + \underline{\hspace{1.5cm}} = 7$	$3 + \underline{\hspace{1.5cm}} = 6$	$3 + \underline{\hspace{1.5cm}} = 5$

6. This is a new way to write addition! The answer goes *under* the line.

a.
$$\begin{array}{r} 4 \\ + \ 3 \\ \hline 7 \end{array}$$

b.
$$\begin{array}{r} 1 \\ + \ 5 \\ \hline \end{array}$$

c.
$$\begin{array}{r} 5 \\ + \ 0 \\ \hline \end{array}$$

d.
$$\begin{array}{r} 4 \\ + \ 1 \\ \hline \end{array}$$

e.
$$\begin{array}{r} 4 \\ + \ 0 \\ \hline \end{array}$$

f.
$$\begin{array}{r} 2 \\ + \ 5 \\ \hline \end{array}$$

g.
$$\begin{array}{r} 0 \\ + \ 3 \\ \hline \end{array}$$

h.
$$\begin{array}{r} 1 \\ + \ 3 \\ \hline \end{array}$$

i.
$$\begin{array}{r} 3 \\ + \ 3 \\ \hline \end{array}$$

j.
$$\begin{array}{r} 2 \\ + \ 2 \\ \hline \end{array}$$

k.
$$\begin{array}{r} 4 \\ + \ 2 \\ \hline \end{array}$$

l.
$$\begin{array}{r} 2 \\ + \ 0 \\ \hline \end{array}$$

m.
$$\begin{array}{r} 1 \\ + \ 6 \\ \hline \end{array}$$

n.
$$\begin{array}{r} 3 \\ + \ 4 \\ \hline \end{array}$$

o.
$$\begin{array}{r} 2 \\ + \ 4 \\ \hline \end{array}$$

7. Solve the word problems. Draw pictures to help you!
   Think: Are you asked the total? Or do you already know the total?

a. Lisa has three goldfish and Lauren has six. How many goldfish do they have together?

b. Paul has seven T-shirts. Two of them are red. How many are not red?

c. A fish bowl has four fish swimming in it. Lisa added four more. How many fish are now in it?

d. Paul has nine toy cars. Six of them are in the living room. The rest of them Paul cannot find. How many cars are missing?

e. Jill wants to have hats for all seven of her dolls. She has found three hats so far. How many does she still need?

f. Brenda ate two cookies, and later she ate four more. How many cookies did she eat?

Puzzle Corner

What numbers can go into these puzzles?

	+		= 6
+		+	
	+		= 5
=		=	
5		6	

	+		= 7
+		+	
	+		= 6
=		=	
7		6	

34

Sums with 8

1. Here are some different ways to group eight marbles into two groups. Write the additions.

_____ + _____ = _____			_____ + _____ = _____
_____ + _____ = _____			_____ + _____ = _____
_____ + _____ = _____			_____ + _____ = _____
_____ + _____ = _____			_____ + _____ = _____

_____ + _____ = _____

2. **Drill.** Do not write the answers here. Just solve them in your head.

☐ + 5 = 8	☐ + 4 = 8	2 + ☐ = 8	3 + ☐ = 8	8 + ☐ = 8
☐ + 0 = 8	☐ + 6 = 8	5 + ☐ = 8	7 + ☐ = 8	6 + ☐ = 8
☐ + 2 = 8	☐ + 3 = 8	1 + ☐ = 8	4 + ☐ = 8	☐ + 1 = 8

3. Play "8 Out" *and/or* "Some Went Hiding" with 8 objects (see the introduction).

35

4. Fill in the missing numbers. You may draw dots to help. Notice the patterns!

a.

$1 + \rule{2cm}{0.4pt} = 8$

$2 + \rule{2cm}{0.4pt} = 8$

$3 + \rule{2cm}{0.4pt} = 8$

$4 + \rule{2cm}{0.4pt} = 8$

b.

$1 + \rule{2cm}{0.4pt} = 7$

$2 + \rule{2cm}{0.4pt} = 7$

$3 + \rule{2cm}{0.4pt} = 7$

$4 + \rule{2cm}{0.4pt} = 7$

c.

$1 + \rule{2cm}{0.4pt} = 6$

$2 + \rule{2cm}{0.4pt} = 6$

$3 + \rule{2cm}{0.4pt} = 6$

$4 + \rule{2cm}{0.4pt} = 6$

5. Draw the missing marbles. Write the additions.

a. $\rule{2cm}{0.4pt} + \rule{2cm}{0.4pt} = 6$

b. $\rule{2cm}{0.4pt} + \rule{2cm}{0.4pt} = 5$

c. $\rule{2cm}{0.4pt} + \rule{2cm}{0.4pt} = 6$

d. $\rule{2cm}{0.4pt} + \rule{2cm}{0.4pt} = 8$

e. $\rule{2cm}{0.4pt} + \rule{2cm}{0.4pt} = 7$

f. $\rule{2cm}{0.4pt} + \rule{2cm}{0.4pt} = 4$

g. $\rule{2cm}{0.4pt} + \rule{2cm}{0.4pt} = 8$

h. $\rule{2cm}{0.4pt} + \rule{2cm}{0.4pt} = 8$

6. Find the missing numbers.

a.	b.	c.	d.
$3 + 4 = \underline{}$	$6 + 2 = \underline{}$	$6 + 1 = \underline{}$	$2 + 5 = \underline{}$
$4 + 4 = \underline{}$	$5 + 2 = \underline{}$	$1 + 7 = \underline{}$	$2 + 6 = \underline{}$
e.	**f.**	**g.**	**h.**
$5 + \underline{} = 7$	$4 + \underline{} = 8$	$3 + \underline{} = 7$	$2 + \underline{} = 8$
$5 + \underline{} = 8$	$4 + \underline{} = 7$	$3 + \underline{} = 8$	$2 + \underline{} = 7$

7. Add.

a. $\begin{array}{r} 4 \\ + 2 \\ \hline \end{array}$
b. $\begin{array}{r} 6 \\ + 2 \\ \hline \end{array}$
c. $\begin{array}{r} 3 \\ + 3 \\ \hline \end{array}$
d. $\begin{array}{r} 7 \\ + 1 \\ \hline \end{array}$
e. $\begin{array}{r} 5 \\ + 2 \\ \hline \end{array}$

f. $\begin{array}{r} 1 \\ + 2 \\ \hline \end{array}$
g. $\begin{array}{r} 6 \\ + 1 \\ \hline \end{array}$
h. $\begin{array}{r} 4 \\ + 3 \\ \hline \end{array}$
i. $\begin{array}{r} 5 \\ + 1 \\ \hline \end{array}$
j. $\begin{array}{r} 3 \\ + 2 \\ \hline \end{array}$

8. Which number is greater? Or are they equal? Write $<$, $>$ or $=$.

a. 7 $\boxed{=}$ 7
b. 7 $\boxed{}$ 8
c. 6 $\boxed{}$ 4
d. 10 $\boxed{}$ 10

e. 8 $\boxed{}$ 4
f. 2 $\boxed{}$ 2
g. 0 $\boxed{}$ 0
h. 8 $\boxed{}$ 7

i. 4 $\boxed{}$ 4
j. 1 $\boxed{}$ 5
k. 6 $\boxed{}$ 8
l. 2 $\boxed{}$ 0

Adding Many Numbers

When you add three numbers, you can add them in any order you wish.

You can add the first two numbers first:	Or you can add the last two numbers first:	Or you can add the first and the last number first:
$1 + 5 + 2$	$1 + 5 + 2$	$1 + 5 + 2$
$6 + 2 = 8$	$1 + 7 = 8$	$5 + 3 = 8$

It doesn't matter in which order you add three numbers.

1. Add in different orders. Which way is easier for you?

a.	b.	c.
Add the last two numbers first.	Add the first two numbers first.	Add the last two numbers first.
$2 + 5 + 2 = $ _____	$3 + 1 + 5 = $ _____	$2 + 5 + 3 = $ _____
Add the first and last number first.	Add the first and last number first.	Add the first two numbers first.
$2 + 5 + 2 = $ _____	$3 + 1 + 5 = $ _____	$2 + 5 + 3 = $ _____

2. Add. Again, you can add in any order.

a.
$$\begin{array}{r} 5 \\ 1 \\ +\ 4 \\ \hline \end{array}$$

b.
$$\begin{array}{r} 3 \\ 1 \\ +\ 4 \\ \hline \end{array}$$

c.
$$\begin{array}{r} 2 \\ 2 \\ +\ 4 \\ \hline \end{array}$$

d.
$$\begin{array}{r} 7 \\ 1 \\ +\ 1 \\ \hline \end{array}$$

e.
$$\begin{array}{r} 4 \\ 0 \\ +\ 3 \\ \hline \end{array}$$

f.
$$\begin{array}{r} 2 \\ 6 \\ +\ 2 \\ \hline \end{array}$$

g.
$$\begin{array}{r} 1 \\ 1 \\ +\ 8 \\ \hline \end{array}$$

h.
$$\begin{array}{r} 3 \\ 2 \\ +\ 4 \\ \hline \end{array}$$

i.
$$\begin{array}{r} 7 \\ 2 \\ +\ 1 \\ \hline \end{array}$$

j.
$$\begin{array}{r} 4 \\ 1 \\ +\ 4 \\ \hline \end{array}$$

3. Solve. You can draw pictures to help.

a. Molly was picking flowers. First she picked two pretty ones.
Then she found some more and picked three more flowers.
Then she picked two more.
How many flowers does Molly have now?

b. Emily put three chairs in a row. Behind them she put another
three chairs, and yet behind them three more chairs.
Draw a picture.
How many chairs did she use?

c. Jack has 10 rabbits. One morning when he came to see them,
he only saw 6 rabbits. How many were missing?

4. Are these additions right? Circle <u>true</u> or <u>false</u>.

a. $1 + 2 + 3 = 6$ true *or* false	**b.** $2 + 2 + 3 = 8$ true *or* false
$1 + 7 + 2 = 9$ true *or* false	$2 + 5 + 2 = 9$ true *or* false

5. Match the addition problems to the right pictures and solve them.

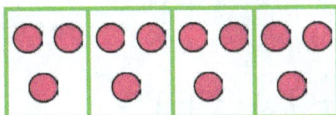

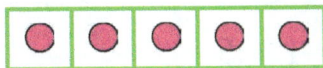

a. $3 + 3 + 3 =$ _____

b. $2 + 2 + 2 + 2 =$ _____

c. $4 + 4 + 4 =$ _____

d. $2 + 2 + 2 =$ _____

e. $3 + 3 + 3 + 3 =$ _____

f. $1 + 1 + 1 + 1 + 1 =$ _____

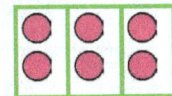

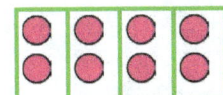

6. Write the additions that match the number line jumps.

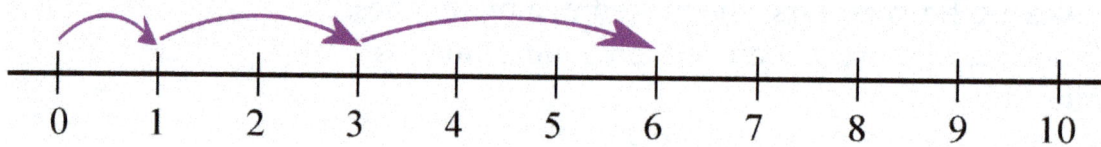

a. _____ + _____ + _____ = _____

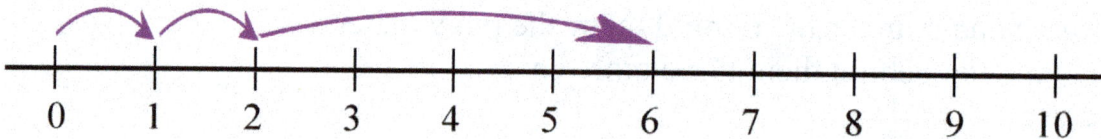

b. _____ + _____ + _____ = _____

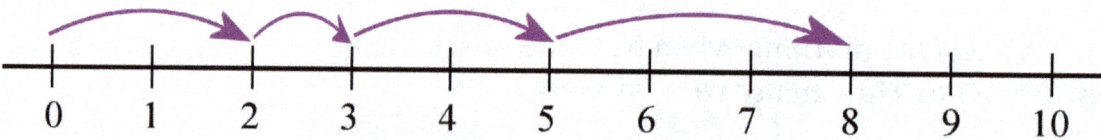

c. _____ + _____ + _____ + _____ = _____

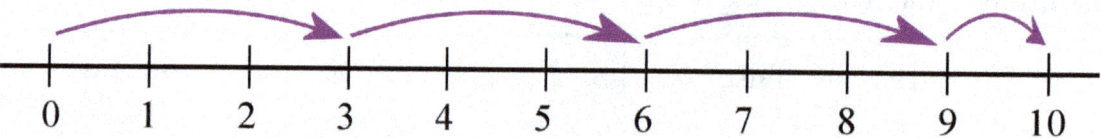

d. _____ + _____ + _____ + _____ = _____

7. Add four numbers. You can color the numbers you want to add first!

a.	b.	c.
$1 + 2 + 2 + 3 =$ _____	$4 + 0 + 3 + 2 =$ _____	$2 + 5 + 3 + 0 =$ _____
$5 + 0 + 1 + 2 =$ _____	$3 + 1 + 2 + 1 =$ _____	$1 + 1 + 2 + 1 =$ _____
$2 + 1 + 3 + 4 =$ _____	$7 + 1 + 1 + 1 =$ _____	$2 + 1 + 5 + 2 =$ _____

Addition Practice 2

1. Add.

a. $4 + 4 =$ _____	**b.** $4 + 3 =$ _____	**c.** $2 + 4 =$ _____
$6 + 2 =$ _____	$5 + 2 =$ _____	$1 + 6 =$ _____

2. **Double** means two times the same thing! Draw dots or sticks. Write the total in the box.

‖ ‖ $\boxed{4}$ **a.** Double 2	**b.** Double 3 $\Box$	**c.** Double 4 $\Box$
d. Double 5 $\Box$	**e.** Double 0 $\Box$	**f.** Double 1 $\Box$

3. Draw jumps for each of the additions. Find the answer.

a. $4 + 2 =$ _____

0 1 2 3 4 5 6 7 8 9 10

b. $6 + 1 =$ _____

0 1 2 3 4 5 6 7 8 9 10

c. $7 + 3 =$ _____

0 1 2 3 4 5 6 7 8 9 10

d. $3 + 6 =$ _____

0 1 2 3 4 5 6 7 8 9 10

4. You can add the numbers in either order! Which way is easier?

a.	b.	c.	d.
7 + 2 = _____	2 + 5 = _____	6 + 2 = _____	1 + 4 = _____
2 + 7 = _____	5 + 2 = _____	2 + 6 = _____	4 + 1 = _____

5. Let's make charts! In the first chart, add one each time. In the second, add two each time. In the third, add three each time.

a.
Add 1

5 + 1 = __6__

6 + 1 = _____

7 + 1 = _____

8 + 1 = _____

9 + 1 = _____

b.
Add 2

2 + 2 = _____

3 + 2 = _____

4 + 2 = _____

5 + 2 = _____

6 + 2 = _____

c.
Add 3

2 + 3 = _____

3 + 3 = _____

4 + 3 = _____

5 + 3 = _____

6 + 3 = _____

6. Fill in the addition tables. Add the number above and the number to the left.

+	1	2	3
1			
2		**4**	
3			

+	1	2	3
4			
5			
6			

Sums with 9

1. Here are some different ways to group nine marbles into two groups. Write the addition sentences.

_____ + _____ = _____		_____ + _____ = _____
_____ + _____ = _____		_____ + _____ = _____
_____ + _____ = _____		_____ + _____ = _____
_____ + _____ = _____		_____ + _____ = _____
_____ + _____ = _____		_____ + _____ = _____

2. **Drill.** Do not write the answers here. Just solve the answers in your head.

$\square + 8 = 9$ $\square + 4 = 9$ $2 + \square = 9$ $3 + \square = 9$ $7 + \square = 9$

$\square + 2 = 9$ $\square + 6 = 9$ $9 + \square = 9$ $6 + \square = 9$ $\square + 1 = 9$

$\square + 7 = 9$ $\square + 3 = 9$ $0 + \square = 9$ $4 + \square = 9$ $\square + 5 = 9$

3. Play "9 Out" *and/or* "Some Went Hiding" with 9 objects (see the introduction).

4. Fill in the missing numbers.

a. △ + 0 = 9

b. 6 + △ = 9

c. 8 + △ = 9

d. △ + 4 = 9

e. 2 + △ = 9

f. 3 + △ = 9

5. Fill in the missing numbers. You may draw dots to help. Notice the patterns!

a.

1 + _____ = 7

2 + _____ = 7

3 + _____ = 7

4 + _____ = 7

b.

1 + _____ = 8

2 + _____ = 8

3 + _____ = 8

4 + _____ = 8

c.

1 + _____ = 9

2 + _____ = 9

3 + _____ = 9

4 + _____ = 9

6. Add.

a.
$$2 + 5$$

b.
$$1 + 6$$

c.
$$4 + 4$$

d.
$$7 + 1$$

e.
$$7 + 2$$

f.
$$3 + 5$$

g.
$$4 + 2$$

h.
$$3 + 4$$

i.
$$1 + 5$$

j.
$$4 + 5$$

44

7. Solve the word problems. Write an addition sentence or a "missing addend" sentence for each problem. Think: "Is it *asking* the total? Or, do I *already know* the total, and something else is being asked?" You can draw a picture to help!

a. Mom has two eggs at home. The cake recipe calls for five eggs. How many more eggs will she need?	**b.** You see four crayons in the crayon box and the rest of them are missing. The full box has eight crayons. How many crayons are missing?
c. Jenny and Penny each have five goldfish. How many do they have together? Betty has three goldfish. How many do the three girls have together?	**d.** You have two dollars. Can you buy a doll for nine dollars? Father has eight dollars. How much money do you have together? Can you buy the doll together?
e. There are two red chairs in the living room and six red chairs in the kitchen, and none in the other rooms. How many red chairs are in the house?	**f.** Joshua has $5. He wants to buy a truck for $7. How many more dollars will he need?
g. If you have $8, and a gift for Mom costs $10, how much more money do you need?	**h.** Jack bought nails for five dollars and screws for four dollars. How much money did he spend in all?

8. First add. Write the answer below (not in the box!). Then write $<$, $>$ or $=$.

a. $5 + 2$ ☐ 4	**b.** $4 + 4$ ☐ 7	**c.** 2 ☐ $1 + 1$	**d.** 7 ☐ $3 + 6$
↓ ↓	↓ ↓	↓ ↓	↓ ↓
☐ 4	☐ 7	☐ ☐	☐ ☐

9. Add in your head. Compare the sum to the other number. Then write $<$ or $>$.

a. $1 + 4$ ☐ 3   **b.** $2 + 2$ ☐ 5   **c.** 2 ☐ $0 + 0$ **d.** 7 ☐ $5 + 3$

e. $4 + 4$ ☐ 9   **f.** $3 + 5$ ☐ 6   **g.** 7 ☐ $6 + 2$ **h.** 8 ☐ $3 + 4$

Puzzle Corner   Complete these number puzzles. Is there more than one solution?

	+		$= 9$
+	▦	+	
	+		$= 9$
$=$		$=$	
10		8	

	+		$= 9$
+	▦	+	
	+		$= 8$
$=$		$=$	
9		8	

Sums with 10

1. Here are some different ways to group ten marbles into two groups. Write the additions.

_____ + _____ = _____

_____ + _____ = _____

_____ + _____ = _____

_____ + _____ = _____

_____ + _____ = _____

_____ + _____ = _____

_____ + _____ = _____

_____ + _____ = _____

_____ + _____ = _____

_____ + _____ = _____

_____ + _____ = _____

2. Play "10 Out" *and/or* "Some Went Hiding" with 10 objects (see the introduction).

47

3. **Drill.** Do not write the answers here. Just think of the answers in your head.

$\square$ + 6 = 10 $\square$ + 4 = 10   1 + $\square$ = 10   6 + $\square$ = 10   3 + $\square$ = 10

$\square$ + 3 = 10 $\square$ + 5 = 10   7 + $\square$ = 10   9 + $\square$ = 10   4 + $\square$ = 10

$\square$ + 8 = 10 $\square$ + 9 = 10   2 + $\square$ = 10   5 + $\square$ = 10   8 + $\square$ = 10

4. Fill in the missing numbers. You may draw dots to help. Notice the patterns!

a.

2 + _____ = 10

3 + _____ = 10

4 + _____ = 10

5 + _____ = 10

b.

2 + _____ = 9

3 + _____ = 9

4 + _____ = 9

5 + _____ = 9

c.

2 + _____ = 8

3 + _____ = 8

4 + _____ = 8

5 + _____ = 8

5. Connect two numbers together if they make ten.

4	1	6	4	4	3
5	2	3	8	2	2
3	7	1	9	2	5
3	6	5	7	5	7
9	0	3	2	3	8

6. Which number is greater? Or are they equal? Write **<** , **>** or **=** .
   (Write one of the alligator mouths or the equal sign).

   a. 6 ☐ 7   b. 10 ☐ 8   c. 6 ☐ 8   d. 10 ☐ 10

   e. 8 ☐ 6   f. 5 ☐ 5   g. 9 ☐ 8   h. 5 ☐ 10

7. First add. Think of the answers in your head. Then compare and write **<** , **>** or **=** .

a. $1 + 9$ ☐ 9	b. $4 + 4$ ☐ 9	c. 6 ☐ $5 + 2$	d. 9 ☐ $5 + 4$
e. $5 + 5$ ☐ 10	f. $3 + 5$ ☐ 7	g. 10 ☐ $6 + 3$	h. 7 ☐ $7 + 1$

8. Which numbers add up to ten? Fill in the missing numbers.

a. ☐ $+ 10 = 10$	b. $6 +$ ☐ $= 10$	c. ☐ $+ 3 = 10$
☐ $+ 5 = 10$	$2 +$ ☐ $= 10$	☐ $+ 8 = 10$
☐ $+ 1 = 10$	$4 +$ ☐ $= 10$	☐ $+ 9 = 10$

9. Draw a line to the correct answer.

	$7 + 1$	
	$2 + 6$	
	$3 + 4$	
7	$5 + 2$	8
	$4 + 4$	
	$1 + 6$	
	$5 + 3$	

	$7 + 3$	
	$3 + 6$	
	$4 + 6$	
9	$1 + 8$	10
	$5 + 4$	
	$3 + 7$	
	$2 + 8$	

10. Solve the word problems. Think: Are you asked the total?
    Or do you already know the total?

a. There were three birds in the tree. Seven more flew in. How many birds are now in the tree?

b. Tina has seven books from the library. She has read three. How many books has she not read?

c. Jessica has ten dolls. She sees four of them in her room. How many are somewhere else?

d. Larry has three toy cars and his brother also has three. How many do they have together?

e. Bill has ten toy cars but he can find only six. How many are missing?

f. Jack saw two birds on the lawn and five on the fence. How many birds did he see in all?

g. Together, Jessica and Jenny have ten books. Jenny has five of them. How many does Jessica have?

h. The store has ten dolls. Two of them are on the bottom shelf. The rest are on the top shelf. How many dolls are on the top shelf?

Comparisons

$7 = 7$ Seven <u>equals</u> seven.	$6 = 2 + 4$ Six <u>equals</u> two plus four.	= means "EQUALS"
$7 < 8$ Seven is <u>less than</u> eight.	$3 + 4 > 5$ Three plus four is <u>greater than</u> five.	< means "IS LESS THAN" > means "IS GREATER THAN"

1. First add in your head. Then write $<$, $>$ or $=$.

a. $4 + 1 \boxed{} 5$	**b.** $7 \boxed{} 4 + 4$	**c.** $6 \boxed{} 2 + 3$
d. $2 + 5 \boxed{} 7$	**e.** $5 \boxed{} 5 + 0$	**f.** $10 \boxed{} 5 + 5$
g. $2 + 2 \boxed{} 3$	**h.** $9 \boxed{} 9$	**i.** $2 \boxed{} 2 + 2$

2. Pick a number to write on the line so the comparison is true.

a. 5 6 7	**b.** 4 5 6	**c.** 5 6 7	**d.** 2 3 4
_____ < 6	_____ < 5	_____ > 6	_____ > 3
e. 9 7 5	**f.** 3 6 9	**g.** 1 3 7	**h.** 2 4 6
_____ > 7	_____ < 5	_____ > 6	_____ < 3

3. Pick a number to write on the line so the comparison is true.

a. 2 3 4	b. 4 5 6	c. 1 2 3
$2 + \underline{\hspace{1.5cm}} = 6$	$1 + \underline{\hspace{1.5cm}} < 6$	$4 + \underline{\hspace{1.5cm}} < 7$
d. 4 5 6	**e.** 4 5 6	**f.** 7 8 9
$2 + \underline{\hspace{1.5cm}} > 6$	$1 + \underline{\hspace{1.5cm}} = 6$	$1 + \underline{\hspace{1.5cm}} > 9$
g. 6 7 8	**h.** 2 4 6	**i.** 4 5 6
$10 = 2 + \underline{\hspace{1.5cm}}$	$3 + \underline{\hspace{1.5cm}} < 7$	$4 + \underline{\hspace{1.5cm}} > 8$

4. Compare. Write $<$, $>$ or $=$.

a. $4 + 3\ \square\ 5$ **b.** $7 + 1\ \square\ 9$ **c.** $4\ \square\ 4 + 2$

d. $2 + 5\ \square\ 8$ **e.** $3 + 4\ \square\ 6$ **f.** $6\ \square\ 3 + 3$

g. $8 + 2\ \square\ 10$ **h.** $9 + 2\ \square\ 9$ **i.** $2\ \square\ 2 + 1$

5. Challenges! First add in your head. Then write $<$, $>$ or $=$.

a. $7 + 3\ \square\ 2 + 8$ **b.** $1 + 1\ \square\ 1 + 4$ **c.** $4\ \square\ 1 + 4$

d. $5 + 4\ \square\ 4 + 5$ **e.** $2 + 5\ \square\ 2 + 2$ **f.** $3\ \square\ 3 + 1$

g. $2 + 4\ \square\ 2 + 1$ **h.** $10 + 0\ \square\ 0 + 10$ **i.** $0\ \square\ 0 + 0$

6. Are these additions right? Circle true or false.

a. $7 + 3 = 10$ true *or* false	**d.** $7 = 1 + 5$ true *or* false
b. $9 = 5 + 5$ true *or* false	**e.** $2 + 2 = 1 + 3$ true *or* false
c. $2 + 4 = 7$ true *or* false	**f.** $3 + 5 = 7 + 2$ true *or* false

7. What numbers make 10? Draw arrows to illustrate the additions on the number line.

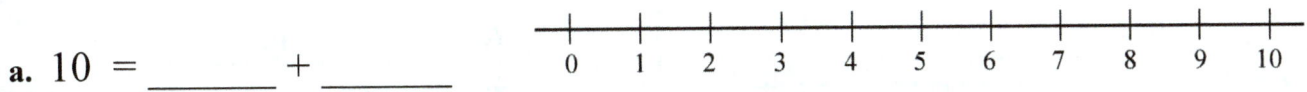

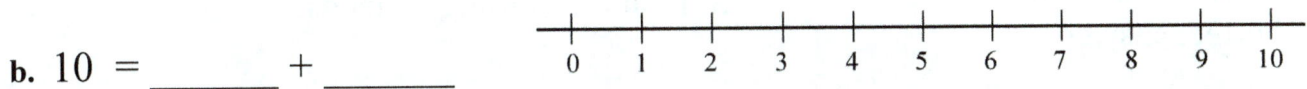

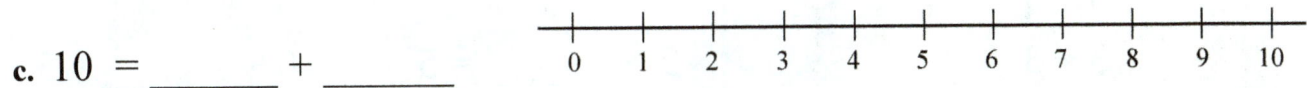

a. $10 =$ _____ $+$ _____

```
 +--+--+--+--+--+--+--+--+--+--+
 0  1  2  3  4  5  6  7  8  9  10
```

b. $10 =$ _____ $+$ _____

```
 +--+--+--+--+--+--+--+--+--+--+
 0  1  2  3  4  5  6  7  8  9  10
```

c. $10 =$ _____ $+$ _____

```
 +--+--+--+--+--+--+--+--+--+--+
 0  1  2  3  4  5  6  7  8  9  10
```

8. Fill in as much of the addition table as you can, and do not worry about the rest.
   Color the square blue if the answer is 8.

+	1	2	3	4	5	6	7
0							
1							
2							
3							
4							
5							

Review of Addition Facts

1. Write different sums of 5 and sums of 6.

$5 =$ _____ $+$ _____

$5 =$ _____ $+$ _____

$5 =$ _____ $+$ _____

$5 =$ _____ $+$ _____

$6 =$ _____ $+$ _____

$6 =$ _____ $+$ _____

$6 =$ _____ $+$ _____

$6 =$ _____ $+$ _____

2. Draw a line to the correct answer.

5

$4 + 1$
$2 + 3$
$3 + 3$
$5 + 0$
$4 + 2$
$5 + 1$
$0 + 6$
$1 + 4$
$2 + 4$

6

3. Find the missing addends.

_____ $+ 2 = 6$ _____ $+ 0 = 6$

$2 +$ _____ $= 5$ $0 +$ _____ $= 5$

$1 +$ _____ $= 5$ $3 +$ _____ $= 6$

$6 +$ _____ $= 6$ $4 +$ _____ $= 6$

_____ $+ 1 = 6$ _____ $+ 4 = 5$

4. Compare. Write $<$, $>$, or $=$.

$2 + 2$ ☐ 5

$2 + 3$ ☐ 5

$2 + 4$ ☐ 5

$4 + 4$ ☐ 5

$5 + 5$ ☐ 5

$5 + 0$ ☐ 5

6 ☐ $2 + 4$

6 ☐ $2 + 5$

6 ☐ $2 + 6$

5. Write different sums of 7 and sums of 8.

7 = _____ + _____	8 = _____ + _____
7 = _____ + _____	8 = _____ + _____
7 = _____ + _____	8 = _____ + _____
7 = _____ + _____	8 = _____ + _____
7 = _____ + _____	8 = _____ + _____
7 = _____ + _____	8 = _____ + _____

6. Draw a line to the correct answer.

7   4 + 3
        2 + 6
        3 + 5
        4 + 4
        5 + 2   8
        1 + 6
        5 + 3
        7 + 1
        6 + 2

7. Find the missing addends

_____ + 2 = 7   _____ + 4 = 8

_____ + 4 = 7   3 + _____ = 7

2 + _____ = 8   3 + _____ = 8

5 + _____ = 8   7 + _____ = 8

6 + _____ = 7   5 + _____ = 7

8. Compare. Write < , > , or = .

3 + 3 ☐ 7   6 + 1 ☐ 7   8 ☐ 6 + 4

4 + 3 ☐ 7   6 + 6 ☐ 7   8 ☐ 4 + 4

5 + 3 ☐ 7   6 + 4 ☐ 7   8 ☐ 5 + 4

9. Write different sums of 9 and sums of 10.

$9 =$ _____ $+$ _____ $10 =$ _____ $+$ _____

$9 =$ _____ $+$ _____ $10 =$ _____ $+$ _____

$9 =$ _____ $+$ _____ $10 =$ _____ $+$ _____

$9 =$ _____ $+$ _____ $10 =$ _____ $+$ _____

$9 =$ _____ $+$ _____ $10 =$ _____ $+$ _____

$9 =$ _____ $+$ _____ $10 =$ _____ $+$ _____

10. Draw a line to the correct answer.

9

$2 + 7$
$3 + 6$
$4 + 6$
$5 + 5$
$9 + 1$
$1 + 8$
$5 + 4$
$3 + 7$
$2 + 8$

10

11. Find the missing addends.

_____ $+ 2 = 10$ _____ $+ 6 = 9$

_____ $+ 4 = 9$ $7 +$ _____ $= 10$

$2 +$ _____ $= 9$ $3 +$ _____ $= 9$

$5 +$ _____ $= 10$ $7 +$ _____ $= 9$

$6 +$ _____ $= 10$ $4 +$ _____ $= 10$

12. Compare. Write $<$, $>$, or $=$.

$2 + 6$ ☐ 9

$4 + 6$ ☐ 9

$3 + 6$ ☐ 9

$6 + 6$ ☐ 10

$5 + 5$ ☐ 10

$4 + 4$ ☐ 10

10 ☐ $10 + 4$

10 ☐ $10 + 0$

10 ☐ $10 + 7$

13. Add.

a.	b.	c.
8 + 1 = _____	4 + 1 + 1 = _____	5 + 2 + 0 + 0 = _____
6 + 2 = _____	8 + 2 + 0 = _____	4 + 3 + 1 + 2 = _____
1 + 7 = _____	1 + 3 + 6 = _____	1 + 2 + 2 + 1 = _____
3 + 4 = _____	2 + 2 + 4 = _____	2 + 3 + 1 + 3 = _____

14. Fill in as much as you can of the addition table.

+	2	4	3	6	7	5	8
1							
3							
4							
2							

Puzzle Corner $\triangle$ represents a number, and $\square$ represents another number. Solve what they are in each case (a, b, and c).
Hint: Make a guess! Then check if your guess is correct.
If not, change your guess.

a.	b.	c.
$\triangle + \triangle = 6$	$\square + \triangle = 7$	$\square + \triangle + \triangle = 7$
$\square + \triangle = 8$	$\square + \square = 10$	$\square + \triangle = 5$

Answer Key

Two Groups and a Total, p. 9

1.

a. 4	b. 4	c. 4
1 and 3	2 and 2	3 and 1
d. 5	e. 5	f. 5
3 and 2	2 and 3	1 and 4

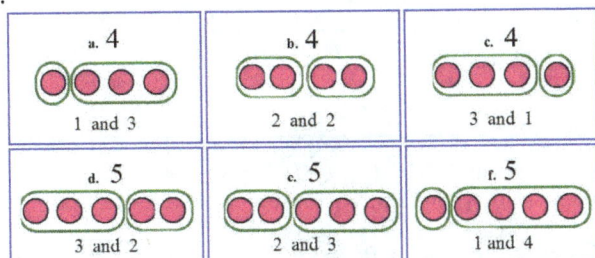

2. a. $1 + 3 = 4$ b. $2 + 2 = 4$ c. $3 + 1 = 4$
   d. $4 + 1 = 5$ e. $3 + 2 = 5$ f. $2 + 3 = 5$
   g. $1 + 4 = 5$ h. $5 + 0 = 5$ i. $0 + 5 = 5$

3. The answers will vary.
   Please check the student's work.

4. a. 1 and 2 b. 4 and 2 c. 2 and 3
   d. 3 and 1 e. 3 and 3 f. 1 and 4

5. a. $2 + 1 = 3$ b. $3 + 1 = 4$ c. $2 + 2 = 4$
   d. $2 + 3 = 5$ e. $1 + 3 = 4$ f. $1 + 1 = 2$
   g. $3 + 2 = 5$ h. $4 + 1 = 5$ i. $1 + 2 = 3$

6. a. 4   b. 4   c. 6   d. 5

Learn the Symbols + and =, p. 12

1. a. 4   b. $1 + 2 = 3$
   c. $3 + 2 = 5$ d. $1 + 4 = 5$
   e. $2 + 3 = 5$ f. $1 + 1 = 2$
   g. $2 + 2 = 4$ h. $3 + 2 = 5$
   i. $3 + 1 = 4$ j. $2 + 2 = 4$

2. a. $1 + 3 = 4$ b. $3 + 2 = 5$
   c. $2 + 3 = 5$ d. $2 + 1 = 3$

3. a. $2 + 0 = 2$ b. $3 + 0 = 3$ c. $4 + 0 = 4$
   d. $0 + 2 = 2$ e. $0 + 5 = 5$ f. $1 + 0 = 1$
   g. $0 + 3 = 3$ h. $0 + 0 = 0$

4. a. $1 + 3 = 4$ b. $2 + 2 = 4$ c. $4 + 1 = 5$
   d. $2 + 0 = 2$ e. $3 + 2 = 5$ f. $0 + 1 = 1$
   g. $2 + 1 = 3$ h. $3 + 0 = 3$ i. $1 + 1 = 2$
   j. $2 + 3 = 5$

Addition Practice 1, p. 15

1. a. $2 + 1 = 3$ b. $3 + 2 = 5$ c. $1 + 2 = 3$
   d. $4 + 1 = 5$ e. $2 + 3 = 5$ f. $0 + 4 = 4$
   g. $2 + 2 = 4$ h. $1 + 0 = 1$ i. $3 + 1 = 4$

2. a. $2 + 2 = 4$ b. $1 + 3 = 4$ c. $0 + 5 = 5$
   d. $4 + 1 = 5$ e. $2 + 3 = 5$ f. $1 + 3 = 4$

3. a. $1 + 2 = 3$ b. $3 + 0 = 3$ c. $2 + 2 = 4$
   d. $2 + 3 = 5$ e. $1 + 4 = 5$ f. $0 + 5 = 5$
   g. $3 + 2 = 5$ h. $2 + 1 = 3$ i. $4 + 1 = 5$

4. b. $1 + 2 = 3$, $2 + 1 = 3$ c. $3 + 1 = 4$, $1 + 3 = 4$
   d. $1 + 4 = 5$, $4 + 1 = 5$ e. $0 + 2 = 2$, $2 + 0 = 2$
   f. $5 + 0 = 5$, $0 + 5 = 5$

Which is More?, p. 17

2. a. 3   b. 5   c. 5
   d. 6; 6 is greater than 2.
    e. 4; 4 is greater than 1.
    f. 4; 2 is less than 4.

3. a. 6   b. 4   c. 5   d. 4
   e. 2   f. 2   g. 5   h. 4

4. a. $1 < 4$ b. $2 < 5$ c. $6 > 3$
   d. $3 < 4$ e. $5 > 1$ f. $2 < 3$

5. a. $1 < 4$ b. $4 > 3$ c. $2 < 5$ d. $0 < 4$

6. a. $1 < 4$ b. $4 < 5$ c. $2 < 4$ d. $5 > 3$
   e. $1 < 2$ f. $3 > 1$ g. $5 > 4$ h. $4 < 6$
   i. $3 < 5$ j. $1 > 0$ k. $2 < 5$ l. $0 < 2$

Missing Items, p. 19

1. a. 2   b. 1   c. 1   d. 2   e. 3
   f. 1   g. 0   h. 3   i. 2

2. a. 2   b. 0   c. 4   d. 2   e. 3   f. 1
   g. 0   h. 2   i. 3   j. 4   k. 0   l. 1

3. a. 2   b. 0   c. 1   d. 3   e. 2   f. 1

4. a. 4   b. 0   c. 3   d. 0   e. 2   f. 1

5. a. 1   b. 1   c. 2   d. 1   e. 1
   f. 2   g. 4   h. 3   i. 0

6. a. 3   b. 2   c. 2   d. 0   e. 2   f. 2   g. 1   h. 2   i. 4

7. a. 2, 3   b. 4, 4   c. 5, 4   d. 5, 5   e. 5, 3   f. 5, 5

8. a. 3, 2, 1, 0   b. 4, 3, 2, 1, 0   c. 5, 4, 3, 2, 1, 0

Sums with 5, p. 24

1.

$0 + 5 = 5$	$5 + 0 = 5$
$1 + 4 = 5$	$4 + 1 = 5$
$2 + 3 = 5$	$3 + 2 = 5$

2.

a. $4 + 1 = 5$	b. $2 + 3 = 5$	c. $1 + 1 = 2$
$2 + 2 = 4$	$1 + 3 = 4$	$0 + 5 = 5$
$3 + 2 = 5$	$1 + 4 = 5$	$1 + 4 = 5$
$1 + 2 = 3$	$2 + 1 = 3$	$3 + 2 = 5$

4. $1 + \underline{4} = 5$ $4 + \underline{1} = 5$ $\underline{3} + 2 = 5$ $\underline{2} + 3 = 5$
   $2 + \underline{3} = 5$ $3 + \underline{2} = 5$ $\underline{5} + 0 = 5$ $\underline{4} + 1 = 5$
   $0 + \underline{5} = 5$ $5 + \underline{0} = 5$ $\underline{1} + 4 = 5$ $\underline{0} + 5 = 5$

5. a. 4, 5, 6   b. 6, 7, 8   c. 3, 4, 5
   d. 7, 8, 9   e. 5, 6, 7   f. 8, 9, 10

6. a. $2 + 3 = 5$ b. $1 + 2 = 3$ c. $3 + 1 = 4$
   d. $4 + 1 = 5$ e. $3 + 3 = 6$ f. $2 + 4 = 6$

Sums with 6, p. 26

1.

$0 + 6 = 6$	$6 + 0 = 6$
$1 + 5 = 6$	$5 + 1 = 6$
$2 + 4 = 6$	$4 + 2 = 6$
$3 + 3 = 6$	

3.

$1 + \underline{5} = 6$ $4 + \underline{2} = 6$ $\underline{4} + 2 = 6$ $\underline{3} + 3 = 6$
$2 + \underline{4} = 6$ $3 + \underline{3} = 6$ $\underline{6} + 0 = 6$ $\underline{5} + 1 = 6$
$6 + \underline{0} = 6$ $5 + \underline{1} = 6$ $\underline{2} + 4 = 6$ $\underline{1} + 5 = 6$

4. a. 6   b. 5   c. 6

5. a. $2 + 4 = 6$ b. $2 + 3 = 5$ c. $4 + 2 = 6$
   d. $3 + 3 = 6$ e. $1 + 5 = 6$ f. $5 + 1 = 6$
   g. $1 + 4 = 5$ h. $0 + 6 = 6$ i. $3 + 2 = 5$

6. a. 3, 4, 0, 2, 5, 1   b. 5, 2, 1, 6, 4, 3

7. 5, 5, 6, 6, 4, 6, 4, 6

Adding on a Number Line, p. 28

1. a. 7

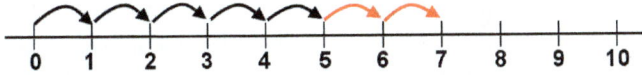

b. 5

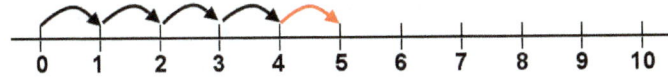

c. 9

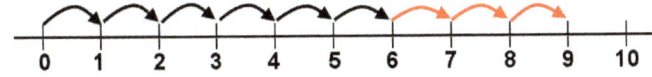

d. 10

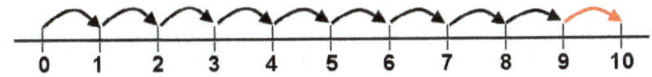

e. 10

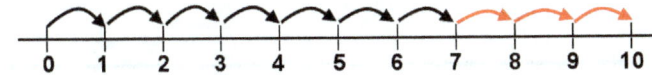

f. 7

2. a. $3 + 3 = 6$ b. $5 + 4 = 9$ c. $2 + 8 = 10$
   d. $6 + 1 = 7$ e. $2 + 5 = 7$ f. $4 + 4 = 8$
   g. $6 + 2 = 8$

3. a. 9

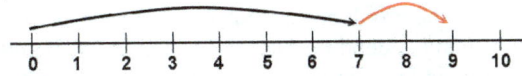

b. 5

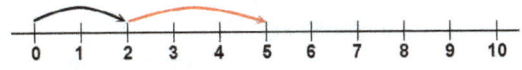

c. 9

,

d. 8

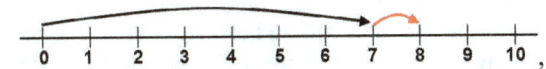
,

e. 10

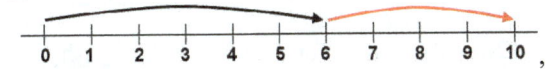
,

f. 6

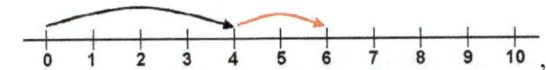
,

g. 10

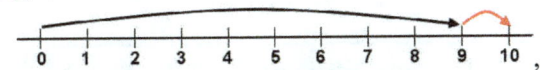
,

h. 5

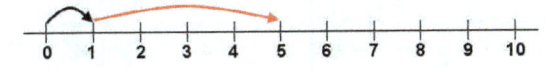

4. a. $8 + 5 = 13$ b. $9 + 3 = 12$ c. $6 + 8 = 14$

5. a. 8, 9   b. 6, 7   c. 7, 8   d. 9, 10
   e. 11,12   f. 13, 14   g. 14, 15   h. 12, 13

Sums with 7, p. 32

1.

$0 + 7 = 7$	$7 + 0 = 7$
$1 + 6 = 7$	$6 + 1 = 7$
$2 + 5 = 7$	$5 + 2 = 7$
$3 + 4 = 7$	$4 + 3 = 7$

2.

$5 + \underline{2} = 7$ $2 + \underline{5} = 7$ $6 + \underline{1} = 7$ $\underline{4} + 3 = 7$ $\underline{0} + 7 = 7$
$3 + \underline{4} = 7$ $1 + \underline{6} = 7$ $0 + \underline{7} = 7$ $\underline{5} + 2 = 7$ $\underline{6} + 1 = 7$
$7 + \underline{0} = 7$ $4 + \underline{3} = 7$ $4 + \underline{3} = 7$ $\underline{1} + 6 = 7$ $\underline{2} + 5 = 7$

3. a. 6, 7   b. 7, 6   c. 7, 7   d. 7, 6

5. a.

$0 + 7 = 7$
$1 + 6 = 7$
$2 + 5 = 7$
$3 + 4 = 7$

b.

$0 + 6 = 6$
$1 + 5 = 6$
$2 + 4 = 6$
$3 + 3 = 6$

c.

$0 + 5 = 5$
$1 + 4 = 5$
$2 + 3 = 5$
$3 + 2 = 5$

6. a. 7   b. 6   c. 5   d. 5   e. 4   f. 7   g. 3   h. 4
   i. 6   j. 4   k. 6   l. 2   m. 7   n. 7   o. 6

7. The use of pictures is optional. It helps many children, though, and in the future—even in algebra word problems—it is a good tactic for solving the problems. Some of the problems are simple addition problems, some are missing addend problems. To distinguish between addition and missing addend problems, you can ask: is the problem *asking* for the total, or do you *already know* the total? Is the problem asking how many there are together, or is it asking how many are missing?

   a. The problem asks for a total. The addition sentence is $3 + 6 = $ <u>9 goldfish.</u>

   b. We know the total is 7. The picture would show initially two shirts, and then the child would draw some more so the total would be seven shirts. The red and other colors together make 7. The addition sentence for this problem is simply $2 + 5 = 7$. <u>Five of the T-shirts are not red.</u>

   c. The problem asks for a total. $4 + 4 = $ <u>8 fish.</u>

7. d. We know the total is 9. The picture would show six toy cars in the living room (perhaps inside a box). Paul has nine. The addition sentence is $6 + 3 = 9$. <u>Three cars are missing.</u>

   e. We know the total is 7. The picture would have seven dolls and three hats. $3 + 4 = 7$. <u>She needs to find 4 more hats.</u>

   f. The problem asks for a total. $2 + 4 = 6$. <u>She ate 6 cookies.</u>

Puzzle Corner. There are many possible solutions; the ones below are just one possibility.

5	+	1	= 6
+		+	
0	+	5	= 5
=		=	
5		6	

1	+	6	= 7
+		+	
6	+	0	= 6
=		=	
7		6	

Sums with 8, p. 35

1.

$0 + 8 = 8$	$8 + 0 = 8$
$1 + 7 = 8$	$7 + 1 = 8$
$2 + 6 = 8$	$6 + 2 = 8$
$3 + 5 = 8$	$5 + 3 = 8$
$4 + 4 = 8$	

2.

$\underline{3} + 5 = 8$ $\underline{4} + 4 = 8$ $2 + \underline{6} = 8$
$\underline{8} + 0 = 8$ $\underline{2} + 6 = 8$ $5 + \underline{3} = 8$
$\underline{6} + 2 = 8$ $\underline{5} + 3 = 8$ $1 + \underline{7} = 8$

$3 + \underline{5} = 8$ $8 + \underline{0} = 8$
$7 + \underline{1} = 8$ $6 + \underline{2} = 8$
$4 + \underline{4} = 8$ $\underline{7} + 1 = 8$

4. a.

$1 + 7 = 8$
$2 + 6 = 8$
$3 + 5 = 8$
$4 + 4 = 8$

b.

$1 + 6 = 7$
$2 + 5 = 7$
$3 + 4 = 7$
$4 + 3 = 7$

c.

$1 + 5 = 6$
$2 + 4 = 6$
$3 + 3 = 6$
$4 + 2 = 6$

5.

a. $2 + 4 = 6$	b. $2 + 3 = 5$
c. $4 + 2 = 6$	d. $5 + 3 = 8$
e. $3 + 4 = 7$	f. $2 + 2 = 4$
g. $3 + 5 = 8$	h. $2 + 6 = 8$

6.

a.	b.	c.	d.
$3 + 4 = 7$ $4 + 4 = 8$	$6 + 2 = 8$ $5 + 2 = 7$	$6 + 1 = 7$ $1 + 7 = 8$	$2 + 5 = 7$ $2 + 6 = 8$
e.	f.	g.	h.
$5 + 2 = 7$ $5 + 3 = 8$	$4 + 4 = 8$ $4 + 3 = 7$	$3 + 4 = 7$ $3 + 5 = 8$	$2 + 6 = 8$ $2 + 5 = 7$

7.

a. $4 + 2 = 6$ b. $6 + 2 = 8$ c. $3 + 3 = 6$ d. $7 + 1 = 8$ e. $5 + 2 = 7$

f. $1 + 2 = 3$ g. $6 + 1 = 7$ h. $4 + 3 = 7$ i. $5 + 1 = 6$ j. $3 + 2 = 5$

8. a. $7 = 7$ b. $7 < 8$ c. $6 > 4$ d. $10 = 10$

   e. $8 > 4$ f. $2 = 2$ g. $0 = 0$ h. $8 > 7$

   i. $4 = 4$ j. $1 < 5$ k. $6 < 8$ l. $2 > 0$

Adding Many Numbers, p. 38

1. a. 9, 9   b. 9, 9   c. 10, 10

2. a. 10   b. 8   c. 8   d. 9   e. 7
   f. 10   g. 10   h. 9   i. 10   j. 9

3. a. <u>She has 7 flowers.</u> $2 + 3 + 2 = 7$
   b. <u>She used 9 chairs.</u> $3 + 3 + 3 = 9$
   c. <u>Four were missing.</u> $10 - 6 = 4$ or $6 + 4 = 10$.

4. a. true, false   b. false, true

5.

a. $3 + 3 + 3 = 9$
b. $2 + 2 + 2 + 2 = 8$
c. $4 + 4 + 4 = 12$
d. $2 + 2 + 2 = 6$
e. $3 + 3 + 3 + 3 = 12$
f. $1 + 1 + 1 + 1 + 1 = 5$

6. a. $1 + 2 + 3 = 6$
   b. $1 + 1 + 4 = 6$
   c. $2 + 1 + 2 + 3 = 8$
   d. $3 + 3 + 3 + 1 = 10$

7. a. 8, 8, 10   b. 9, 7, 10   c. 10, 5, 10

Addition Practice 2, p. 41

1.

a. $4 + 4 = 8$ $6 + 2 = 8$	b. $4 + 3 = 7$ $5 + 2 = 7$	c. $2 + 4 = 6$ $1 + 6 = 7$

2. a. 4   b. 6   c. 8   d. 10   e. 12   f. 2

3.

a. $4 + 2 = 6$

b. $6 + 1 = 7$

c. $7 + 3 = 10$

d. $3 + 6 = 9$

4. a. 9, 9   b. 7, 7   c. 8, 8   d. 5, 5

5.

a. Add 1	b. Add 2	c. Add 3
$5 + 1 = \underline{6}$	$2 + 2 = 4$	$2 + 3 = 5$
$6 + 1 = 7$	$3 + 2 = 5$	$3 + 3 = 6$
$7 + 1 = 8$	$4 + 2 = 6$	$4 + 3 = 7$
$8 + 1 = 9$	$5 + 2 = 7$	$5 + 3 = 8$
$9 + 1 = 10$	$6 + 2 = 8$	$6 + 3 = 9$

6.

+	1	2	3
1	2	3	4
2	3	4	5
3	4	5	6

+	1	2	3
4	5	6	7
5	6	7	8
6	7	8	9

Sums with 9, p. 43

1.

0 + 9 = 9	9 + 0 = 9
1 + 8 = 9	8 + 1 = 9
2 + 7 = 9	7 + 2 = 9
3 + 6 = 9	6 + 3 = 9
4 + 5 = 9	5 + 4 = 9

2. $\underline{1} + 8 = 9$ $\underline{5} + 4 = 9$ $2 + \underline{7} = 9$ $3 + \underline{6} = 9$ $7 + \underline{2} = 9$
   $\underline{7} + 2 = 9$ $\underline{3} + 6 = 9$ $9 + \underline{0} = 9$ $6 + \underline{3} = 9$ $\underline{8} + 1 = 9$
   $\underline{2} + 7 = 9$ $\underline{6} + 3 = 9$ $0 + \underline{9} = 9$ $4 + \underline{5} = 9$ $\underline{4} + 5 = 9$

4. a. 9   b. 3   c. 1   d. 5   e. 7   f. 6

5.

a.	b.	c.
1 + 6 = 7	1 + 7 = 8	1 + 8 = 9
2 + 5 = 7	2 + 6 = 8	2 + 7 = 9
3 + 4 = 7	3 + 5 = 8	3 + 6 = 9
4 + 3 = 7	4 + 4 = 8	4 + 5 = 9

6. a. 2 + 5 = 7   b. 1 + 6 = 7   c. 4 + 4 = 8   d. 7 + 1 = 8   e. 7 + 2 = 9   f. 3 + 5 = 8   g. 4 + 2 = 6   h. 3 + 4 = 7   i. 1 + 5 = 6   j. 4 + 5 = 9

7. a. We know the total is 5. 2 + 3 = 5. <u>She needs 3 more eggs.</u>
   b. The total is 8. 4 + 4 = 8. So, <u>4 crayons are missing.</u>
   c. The problem asks for the total. 5 + 5 = 10. Jenny and Penny have 10 goldfish. With Betty's fish, there are 10 + 3 = <u>13 goldfish altogether.</u>
   d. You cannot buy the doll with two dollars. 8 + 2 = 10. You have $10 together. You can buy the doll together.
   e. The problem asks for the total. 2 + 6 = 8. <u>There are 8 red chairs in the house.</u>
   f. <u>Two dollars more.</u> 5 + 2 = 7
   g. <u>Two dollars more.</u> 8 + 2 = 10
   h. <u>Nine dollars.</u> 5 + 4 = 9

8.

a.	b.	c.	d.
5 + 2 4	4 + 4 7	2 1 + 1	7 3 + 6
↓ ↓	↓ ↓	↓ ↓	↓ ↓
7 > 4	8 > 7	2 = 2	7 < 9

9. a. 1 + 4 > 3   b. 2 + 2 < 5   c. 2 > 0 + 0   d. 7 < 5 + 3

   e. 4 + 4 < 9   f. 3 + 5 > 6   g. 7 < 6 + 2   h. 8 > 3 + 4

Puzzle Corner. There are many solutions. The ones below are just examples.

1	+	8	= 9
+		+	
9	+	0	= 9
=		=	
10		8	

1	+	8	= 9
+		+	
8	+	0	= 8
=		=	
9		8	

64

1.

🌰🌰🌰🌰🌰🌰🌰🌰🌰🌰	🌰🌰🌰🌰🌰🌰🌰🌰🌰🌰
0 + 10 = 10	10 + 0 = 10
🌰 | 🌰🌰🌰🌰🌰🌰🌰🌰🌰	🌰🌰🌰🌰🌰🌰🌰🌰🌰 | 🌰
1 + 9 = 10	9 + 1 = 10
🌰🌰 | 🌰🌰🌰🌰🌰🌰🌰🌰	🌰🌰🌰🌰🌰🌰🌰🌰 | 🌰🌰
2 + 8 = 10	8 + 2 = 10
🌰🌰🌰 | 🌰🌰🌰🌰🌰🌰🌰	🌰🌰🌰🌰🌰🌰🌰 | 🌰🌰🌰
3 + 7 = 10	7 + 3 = 10
🌰🌰🌰🌰 | 🌰🌰🌰🌰🌰🌰	🌰🌰🌰🌰🌰🌰 | 🌰🌰🌰🌰
4 + 6 = 10	6 + 4 = 10
🌰🌰🌰🌰🌰 | 🌰🌰🌰🌰🌰	
5 + 5 = 10	

3.

4 + 6 = 10	6 + 4 = 10	1 + 9 = 10
7 + 3 = 10	5 + 5 = 10	7 + 3 = 10
2 + 8 = 10	1 + 9 = 10	2 + 8 = 10
6 + 4 = 10	3 + 7 = 10	
9 + 1 = 10	4 + 6 = 10	
5 + 5 = 10	8 + 2 = 10	

4.

a.	b.	c.
▪▪ ▪▪▪▪▪▪▪▪ 2 + 8 = 10	▪▪ ▪▪▪▪▪▪▪ 2 + 7 = 9	▪▪ ▪▪▪▪▪▪ 2 + 6 = 8
▪▪▪ ▪▪▪▪▪▪▪ 3 + 7 = 10	▪▪▪ ▪▪▪▪▪▪ 3 + 6 = 9	▪▪▪ ▪▪▪▪▪ 3 + 5 = 8
▪▪▪▪ ▪▪▪▪▪▪ 4 + 6 = 10	▪▪▪▪ ▪▪▪▪▪ 4 + 5 = 9	▪▪▪▪ ▪▪▪▪ 4 + 4 = 8
▪▪▪▪▪ ▪▪▪▪▪ 5 + 5 = 10	▪▪▪▪▪ ▪▪▪▪ 5 + 4 = 9	▪▪▪▪▪ ▪▪▪ 5 + 3 = 8

5.

6. a. 6 < 7   b. 10 > 8   c. 6 < 8   d. 10 = 10
   e. 8 > 6   f. 5 = 5   g. 9 > 8   h. 5 < 10

7.

a. 1 + 9 > 9	b. 4 + 4 < 9	c. 6 < 5 + 2	d. 9 = 5 + 4
e. 5 + 5 = 10	f. 3 + 5 > 7	g. 10 > 6 + 3	h. 7 < 7 + 1

8.

a. 0 + 10 = 10 5 + 5 = 10 9 + 1 = 10	b. 6 + 4 = 10 2 + 8 = 10 4 + 6 = 10	c. 7 + 3 = 10 2 + 8 = 10 1 + 9 = 10

9.

10. a. The problem asks for the total. 3 + 7 = 10 birds now.
    b. We already know the total is 7. 3 + 4 = 7
       There are four books that she has not read.
    c. We already know the total is 10. 4 + 6 = 10
       Six of the dolls are not in her room.
    d. The problem asks for the total. 3 + 3 = 6;
       They have 6 cars.
    e. We already know the total is 10. 6 + 4 = 10
       4 are missing.
    f. The problem asks for the total. 2 + 5 = 7 birds.
    g. We already know the total is 10. 5 + 5 = 10
       Jessica has 5 books.
    h. We already know the total is 10. 2 + 8 = 10.
       There are 8 dolls on the top shelf.

65

Comparisons, p. 51

1.

a. $4 + 1 = 5$	b. $7 < 4 + 4$	c. $6 > 2 + 3$
d. $2 + 5 = 7$	e. $5 = 5 + 0$	f. $10 = 5 + 5$
g. $2 + 2 > 3$	h. $9 = 9$	i. $2 < 2 + 2$

2.

a. $5 < 6$	b. $4 < 5$	c. $7 > 6$	d. $2 = 2$
e. $7 < 9$	f. $5 > 3$	g. $7 = 7$	h. $3 > 2$

3.

a. $2 + 4 = 6$	b. $1 + 4 < 6$	c. $4 + \underline{1 \text{ or } 2} < 7$
d. $2 + \underline{5 \text{ or } 6} > 6$	e. $1 + 5 = 6$	f. $1 + 9 > 9$
g. $10 = 2 + 8$	h. $3 + 2 < 7$	i. $4 + \underline{5 \text{ or } 6} > 8$

4.

a. $4 + 3 > 5$	b. $7 + 1 < 9$	c. $4 < 4 + 2$
d. $2 + 5 < 8$	e. $3 + 4 > 6$	f. $6 = 3 + 3$
g. $8 + 2 = 10$	h. $9 + 2 > 9$	i. $2 < 2 + 1$

5. a. $7 + 3 = 2 + 8$ b. $1 + 1 < 1 + 4$ c. $4 < 1 + 4$

   d. $5 + 4 = 4 + 5$ e. $2 + 5 > 2 + 2$ f. $3 < 3 + 1$

   g. $2 + 4 > 2 + 1$ h. $10 + 0 = 0 + 10$ i. $0 = 0 + 0$

6.

a. true	d. false
b. false	e. true
c. false	f. false

7. Answers will vary. For example:

   a. $10 = 9 + 1 = 10$

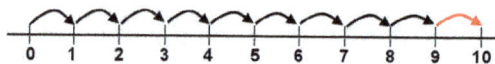

   b. $10 = 6 + 4 = 10$

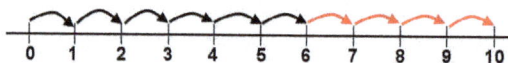

   c. $10 = 5 + 5 = 10$

8.

+	1	2	3	4	5	6	7
0	1	2	3	4	5	6	7
1	2	3	4	5	6	7	8
2	3	4	5	6	7	8	9
3	4	5	6	7	8	9	10
4	5	6	7	8	9	10	11
5	6	7	8	9	10	11	12

Review of Addition Facts, p. 54

1. Answers will vary.

5 = 1 + 4	6 = 0 + 6
5 = 2 + 3	6 = 1 + 5
5 = 0 + 5	6 = 2 + 4
5 = 3 + 2	6 = 3 + 3

2.

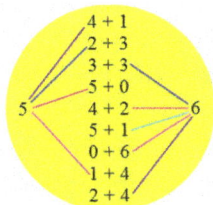

3.

4 + 2 = 6	6 + 0 = 6
2 + 3 = 5	0 + 5 = 5
1 + 4 = 5	3 + 3 = 6
6 + 0 = 6	4 + 2 = 6
5 + 1 = 6	1 + 4 = 5

4.

2 + 2 < 5	4 + 4 > 5	6 = 2 + 4
2 + 3 = 5	5 + 5 > 5	6 < 2 + 5
2 + 4 > 5	5 + 0 = 5	6 < 2 + 6

5. Answers will vary.

7 = 0 + 7	8 = 0 + 8
7 = 1 + 6	8 = 1 + 7
7 = 2 + 5	8 = 2 + 6
7 = 3 + 4	8 = 3 + 5
7 = 4 + 3	8 = 4 + 4
7 = 5 + 2	8 = 5 + 3

6.

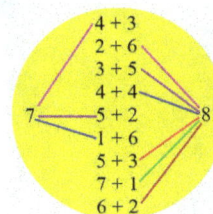

7.

5 + 2 = 7	4 + 4 = 8
3 + 4 = 7	3 + 4 = 7
2 + 6 = 8	3 + 5 = 8
5 + 3 = 8	7 + 1 = 8
6 + 1 = 7	5 + 2 = 7

8.

3 + 3 < 7	6 + 1 = 7	8 < 6 + 4
4 + 3 = 7	6 + 6 > 7	8 = 4 + 4
5 + 3 > 7	6 + 4 > 7	8 < 5 + 4

9. Answers will vary.

9 = 9 + 0	10 = 0 + 10
9 = 1 + 8	10 = 1 + 9
9 = 7 + 2	10 = 2 + 8
9 = 3 + 6	10 = 3 + 7
9 = 4 + 5	10 = 4 + 6
9 = 2 + 7	10 = 5 + 5

10.

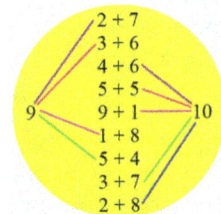

11.

8 + 2 = 10	3 + 6 = 9
5 + 4 = 9	7 + 3 = 10
2 + 7 = 9	3 + 6 = 9
5 + 5 = 10	7 + 2 = 9
6 + 4 = 10	4 + 6 = 10

12.

2 + 6 < 9	6 + 6 > 10	10 < 10 + 4
4 + 6 > 9	5 + 5 = 10	10 = 10 + 0
3 + 6 = 9	4 + 4 < 10	10 < 10 + 7

13.

a.	b.	c.
8 + 1 = 9	4 + 1 + 1 = 6	5 + 2 + 0 + 0 = 7
6 + 2 = 8	8 + 2 + 0 = 10	4 + 3 + 1 + 2 = 10
1 + 7 = 8	1 + 3 + 6 = 10	1 + 2 + 2 + 1 = 6
3 + 4 = 7	2 + 2 + 4 = 8	2 + 3 + 1 + 3 = 9

14.

+	2	4	3	6	7	5	8
1	3	5	4	7	8	6	9
3	5	7	6	9	10	8	11
4	6	8	7	10	11	9	12
2	4	6	5	8	9	7	10

Puzzle corner:

a. △ = 3 and ☐ = 5

b. ☐ = 5 and △ = 2

c. ☐ = 3 and △ = 2

More from math MAMMOTH

Math Mammoth has a variety of resources to fit your needs. All are available as economical downloads, and most also as printed copies.

- **Math Mammoth Light Blue Series**
  A complete curriculum for grades 1-7. Each grade level includes two student worktexts (A and B), which contain all the instruction and exercises all in the same book, answer keys, tests, cumulative reviews, and a worksheet maker. International (all metric), Canadian, and South African versions are also available.

  https://www.MathMammoth.com/complete-curriculum

  https://www.MathMammoth.com/international/international

  https://www.MathMammoth.com/canada/

  https://www.MathMammoth.com/south_africa/

- **Math Mammoth Skills Review Workbooks**
  These workbooks are intended to be used alongside the Light Blue series full curriculum, and they provide additional review to the topics studied in the main curriculum, in a spiral manner.

  https://www.MathMammoth.com/skills_review_workbooks/

- **Math Mammoth Blue Series**
  Blue Series books are topical worktexts for grades 1-8, containing both instruction and exercises. They cover all elementary math topics from 1st through 7th grade and some for 8th grade. These books are not tied to grade levels, and are thus great for filling in gaps.

  https://www.MathMammoth.com/blue-series

- **Make It Real Learning**
  These activity workbooks concentrate on answering the question, "Where is math used in real life?" The series includes various workbooks for grades 3-12.

  https://www.MathMammoth.com/worksheets/mirl/

- **Review Workbooks**
  Workbooks for grades 1-7 that provide a comprehensive review of one grade level of math—for example, for review during school break or summer vacation.

  https://www.MathMammoth.com/review_workbooks/

Free gift!

- Receive over 350 free sample pages and worksheets from my books, plus other freebies:
  https://www.MathMammoth.com/worksheets/free

Lastly...

- Inspire4 is an inspirational website for the whole family I've been privileged to help with:
  https://www.inspire4.com